MATHEMATICAL SURVEYS AND MONOGRAPHS SERIES LIST

BASIC HYPERGEOMETRIC
SERIES AND APPLICATIONS

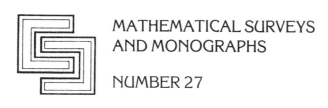

MATHEMATICAL SURVEYS
AND MONOGRAPHS

NUMBER 27

BASIC HYPERGEOMETRIC SERIES AND APPLICATIONS

NATHAN J. FINE

American Mathematical Society
Providence, Rhode Island

1980 *Mathematics Subject Classification* (1985 *Revision*). Primary 05A15, 05A17, 05A19, 05A30, 11E25, 11P57, 11P76, 11P80, 33A70.

LIBRARY OF CONGRESS

Library of Congress Cataloging-in-Publication Data

Fine, Nathan J. (Nathan Jacob), 1916-
  Basic hypergeometric series and applications/Nathan J. Fine.
    p. cm. -- (Mathematical surveys and monographs, ISSN 0076-5376; no. 27)
  Includes bibliographies.
  ISBN 0-8218-1524-5 (alk. paper)
  1. Hypergeometric series. I. Title. II. Series.
QA295.F54   1988
515′.243--dc19   88-6235

Contents

Foreword

In 1948, Nathan Fine published a note in the Proceedings of the National Academy of Sciences announcing several elegant and intriguing partition theorems. These results were marked both by their simplicity of statement and (as we shall see in Chapter 2) by the depth of their proof. Fine was at that time engaged in his own special development of q-hypergeometric series, and as the years passed he kept adding to his results and polishing his presentation. Several times, both at Penn and Penn State, he presented courses on this material. I took the course twice, first in 1962–63 at Penn and then in 1968–69 at Penn State. As a graduate student at Penn, I wrote my thesis on mock theta functions under Rademacher's direction. The material that Fine was lecturing about fit in perfectly with my thesis work and introduced me to many aspects of this extensive subject. The course was truly inspiring. As I look back at it, it is hard for me to decide whether the course material or Fine's exquisite presentation of it impressed me most.

Over the years, Fine's work and the related course notes have greatly assisted me in my work. They were especially helpful in my study of Ramanujan's "Lost" Notebook which overlaps the present book in significant ways.

Research on q-hypergeometric series is significantly more active now than when Fine began his researches. There are now major interactions with Lie algebras, combinatorics, special functions, and number theory.

I am immensely pleased that Fine has finally decided to publish this monograph, and I am grateful to him for allowing me to provide some chapter notes. This book has had a profound impact on my career, and I am glad to see it become available to the mathematical public generally.

George E. Andrews

Introduction

The theory of partitions, founded by Euler, has led in a natural way to series involving factors of the form

$$(1 - aq)(1 - aq^2) \cdots (1 - aq^n).$$

These "basic hypergeometric series" or "Eulerian series" were studied systematically first by Heine [27]. Many early results go back to Euler, Gauss, and Jacobi. A short account is given by Bailey [5], who has made many contributions of his own. Ample references will be found in [3], in Hardy and Wright [25], in MacMahon [30], and in Dickson's *History* [11]. Among the later systematic developments of the theory should be mentioned those of Hahn [20, 21, 22], and Sears [39, 40, 41]. For very complete references and expositions, see Andrews [1, 2].

The subject of basic hypergeometric series has been developed to such an extent, with such a profusion of powerful and general results, expressed in so compact a notation, as to make it appear quite formidable to the neophyte. Nevertheless, the beauty and surprising nearness to the surface of some of the results could easily lead one to embark on an almost uncharted investigation of his own. It was this course that I followed, starting with a modicum of casual information, many, many years ago.

By taking this approach, it was inevitable that I should be rediscovering much that was known to even the earliest workers in the field. Still, it was encouraging that many results obtained in this way appeared to be worthwhile and new, and that old ones dropped out as easy by-products. The present work is an outcome of this exploratory journey, and is a welcome opportunity to share with others my joys of discovery.

At least at the start, I have restricted myself deliberately to the study of a power series in t with coefficients having one Eulerian factor in numerator and denominator. This function,

$$F(a, b; t) = 1 + \frac{(1 - aq)}{(1 - bq)}t + \frac{(1 - aq)(1 - aq^2)}{(1 - bq)(1 - bq^2)}t^2 + \cdots,$$

is a special case of the Heine series. It satisfies first-order linear difference-equations in each of the three arguments, such as

$$(1 - t)F(a, b; t) = (1 - b) + (b - atq)F(a, b; tq).$$

These are given in §2 and §4 of Chapter 1, and in §5 there is presented an elementary method for using them to obtain transformations of F. This method is exploited in §§6–14; the next three sections contain special developments to be used later.

In §18, the series for $F(a, b; t)$ is extended to the left, and the resulting bilateral series is summed as an infinite product. I later found that this elegant and fruitful result goes back to Ramanujan. The exploitation of a special case permits one to evaluate the coefficients of several classes of infinite products in terms of divisor sums (§19) (see also §10). The latter are generally of the form $E_{r,s,...}(N; p)$, denoting the excess of the number of divisors of N congruent to $r, s, \ldots \pmod{p}$ over the number congruent to $-r, -s, \ldots$. §§20 and 21 contain a small start in the direction of a general transformation theory for basic series in many variables. The treatment is still quite elementary, but some of the results seem to be new and interesting. The basic multinomial (§21) is a generalization of a finite sum given in two special cases by Gauss, who used them to evaluate the Gaussian sums.

The developments of Chapter 1 lead naturally, by paraphrase, into the arithmetic domains of partition theory (Chapter 2), theorems of Liouville type (§§28, 29, 30), and sums of squares (§31). Contact is also made with the mock theta-functions of Ramanujan (Chapter 3); these are linked with the rank of partitions, introduced by Dyson [13] and treated by Atkin and Swinnerton-Dyer [3].

In §32 I give a number of examples of modular functions $A + B \sum C(N)q^N$ in which the coefficients $C(N)$ are multiplicative. These are, of course, related to the Hecke theory [26]. In all our cases, the modular functions are products of Dedekind η-functions, of a class defined and studied by Ramanujan (see [23, 24]). As one example, which I regard as very beautiful, I quote (32.5):

$$\prod_{n \geq 1} \frac{(1 - q^{2n})(1 - q^{3n})(1 - q^{8n})(1 - q^{12n})}{(1 - q^n)(1 - q^{24n})} = 1 + \sum E_{1,5,7,11}(N; 24)q^N.$$

Chapter 5 contains the beginnings of an elementary constructive approach to the field of modular equations. §§40, 41, 42 deal with a system of curious identities related to theta-functions.

Over the years I have been assisted greatly by many persons and institutions. Among them, I wish to acknowledge the Office of Naval Research during the year 1949–50, under Contract N9 ONR 90,000, the National Science Foundation when I was a NSF Post-doctoral Fellow in 1953–54, the Guggenheim Foundation when I was a Fellow in 1958–59, the University of Pennsylvania and the Pennsylvania State University when I was on their faculties. Finally, I am tremendously indebted to George Andrews, without whose assistance and forceful encouragement this work would not have seen the light of day. In addition, he has greatly enhanced it with his scholarly Chapter Notes. The references in square brackets in the Notes are to his list of References following each Chapter. Those in the text refer to the Bibliography.

Some of the material was presented by me as a Hedrick Memorial Lecturer in 1966. At that time I was invited to expand it as a Survey and accepted, but was unable to fulfill my commitment because of ill health. The American Mathematical Society was kind enough to revitalize that invitation more recently. For this I am truly grateful.

To Ruth,
Naomi, Emily, and Isabel

CHAPTER 1

Fundamental Properties of Basic Hypergeometric Series

1. Definitions. We propose to study the series

$$(1.1) \qquad F(a,b;t:q) = 1 + \sum_{n\geq 1} \frac{(1-aq)(1-aq^2)\cdots(1-aq^n)}{(1-bq)(1-bq^2)\cdots(1-bq^n)} t^n.$$

In general, q will denote a fixed complex number of absolute value less than 1, so that we may write $q = \exp(\pi i \tau)$, where $I(\tau) > 0$. By q^c we shall always mean $\exp(c\pi i \tau)$. For $n \geq 1$, we define

$$(1.2) \qquad (a)_n = (a;q)_n = (1-a)(1-aq)\cdots(1-aq^{n-1}),$$

and for $n = 0$,

$$(1.21) \qquad (a)_0 = 1.$$

We shall also use

$$(1.22) \qquad (a)_\infty = (a;q)_\infty = \prod_{n\geq 0}(1-aq^n).$$

In general, where there is no danger of ambiguity, the parameter q will be omitted from the notation. Thus (1.1) may be written

$$(1.1') \qquad F(a,b;t) = \sum_{n\geq 0} \frac{(aq)_n}{(bq)_n} t^n.$$

We shall also have occasion to use the Gauss polynomials:

$$(1.3) \qquad \begin{bmatrix} m \\ n \end{bmatrix} = \begin{cases} \dfrac{(q)_m}{(q)_n(q)_{m-n}} & (0 \leq n \leq m), \\ 0 & (0 \leq m < n). \end{cases}$$

That these are indeed polynomials in q is easy to prove, and we shall not stop to do so here. They are a generalization of the binomial coefficients, to which they reduce for $q = 1$.

2. Two functional equations. It is clear from the definition that

$$(2.1) \qquad (aq)_n = (1-aq)(aq^2)_{n-1} \qquad (n \geq 1).$$

Hence (1.1') may be rewritten as

$$(2.2) \qquad F(a,b;t) = 1 + \frac{(1-aq)}{(1-bq)} t F(aq, bq; t).$$

1

This is the first of a number of functional equations that we shall derive and use. To derive the second, write A_n for the coefficient of t^n in (1.1'). Then $A_0 = 1$ and

(2.3) $$(1 - bq^{n+1})A_{n+1} = (1 - aq^{n+1})A_n \qquad (n \geq 0).$$

Multiply (2.3) by t^{n+1} and sum for $n \geq 0$; then

$$\sum_{n \geq 0} A_{n+1}t^{n+1} - b\sum_{n \geq 0} A_{n+1}(tq)^{n+1} = t\sum_{n \geq 0} A_n t^n - atq\sum_{n \geq 0} A_n(tq)^n,$$

which leads easily to the required

(2.4) $$F(a, b; t) = \frac{1 - b}{1 - t} + \frac{b - atq}{1 - t}F(a, b; tq).$$

Equation (2.2) allows us to advance the parameters (a, b) to (aq, bq), and (2.4) advances t to tq. We shall refer to these as the transformations $(a, b) \to (aq, bq)$, $t \to tq$, and so on.

3. The analytic character of $F(a, b; t)$. Thus far we have been operating in a purely formal fashion, and a few remarks on convergence and analyticity are in order. The partial products $(aq)_n$ converge for all values of a, as may easily be seen from the absolute convergence of $\sum q^n$. Hence, if b is not one of the values q^{-1}, $q^{-2}, \ldots$, the coefficients $(aq)_n/(bq)_n$ are bounded, and the series (1.1') converges for all t inside the unit circle, and represents an analytic function there. Hence the function on the right of (2.4) is regular in $|t| < |q|^{-1}$ except (possibly) for a simple pole at $t = 1$. Thus we obtain the continuation of F to the larger circle. It is then easy to apply (2.4) again to continue F to the circle $|t| < |q|^{-2}$, and in this way we reach the conclusion that for $b \neq q^{-n}$, $n \geq 1$, the only possible singularities of F are at the points $t = q^{-n}$, $n \geq 0$, which are simple poles in general. As a function of b, F is regular except possibly at the simple poles $b = q^{-n}$, $n \geq 1$, provided that $t \neq q^{-n}$, $n \geq 0$. Finally, as a function of a, F is entire, provided that b and t do not have one of the singular values mentioned above.

4. More transformations. By combining (2.2) and (2.4) (with a, b replaced by aq, bq in the latter), we obtain easily

(4.1) $$F(a, b; t) = \frac{1 - atq}{1 - t} + \frac{(1 - aq)(b - atq)}{(1 - bq)(1 - t)}tqF(aq, bq; tq).$$

Let us write

(4.2) $$(1 - t)F(a, b; t) = 1 + \sum_{n \geq 1}(A_n - A_{n-1})t^n.$$

Recalling that $A_0 = 1$ and

$$A_{n-1} = \frac{1 - bq^n}{1 - aq^n}A_n \qquad (n \geq 1),$$

we have

$$A_n - A_{n-1} = (b - a)q^n\frac{A_n}{1 - aq^n} \qquad (n \geq 1).$$

Hence, from (4.2)

$$(1-t)F(a,b;t) = 1 + (b-a)\sum_{n\geq 1}\frac{(aq)_{n-1}}{(bq)_n}(tq)^n$$

$$= 1 + \frac{b-a}{1-bq}\sum_{n\geq 1}\frac{(aq)_{n-1}}{(bq^2)_{n-1}}(tq)^n,$$

that is,

(4.3) $$F(a,b;t) = \frac{1}{1-t} + \frac{(b-a)tq}{(1-bq)(1-t)}F(a,bq;tq).$$

Combining (4.3) with (2.4), we obtain

(4.4) $$F(a,b;t) = \frac{b}{b-at} + \frac{(b-a)t}{(1-bq)(b-at)}F(a,bq;t),$$

and from (4.4) and (2.2),

(4.5) $$F(a,b;t) = -\frac{(1-b)aq}{b-aq} + \frac{(1-aq)(b-atq)}{b-aq}F(aq,b;t).$$

Finally, (4.5) and (2.4) yield

(4.6) $$F(a,b;t) = \left(\frac{1-b}{1-t}\right)\left\{1 - \frac{(b-atq)}{(b-aq)}aq\right\}$$
$$+ \frac{(1-aq)(b-atq)(b-atq^2)}{(1-t)(b-aq)}F(aq,b;tq).$$

It is obvious that we can now obtain any transformation of the form $(a,b,t) \to (aq^m, bq^n, tq^p)$, where m, n, p are arbitrary integers.

5. The method of iteration. In this section we indicate an elementary method for solving difference equations of the form

(5.1) $$f_n = L_n + M_n f_{n+1} \qquad (n \geq 0),$$

not for f_n but for f_0. This method will be useful in obtaining new forms for $F(a,b;t)$. It is clear that by use of (5.1) for $n = 0, 1, 2, \ldots, N$, we may eliminate successively $f_1, f_2, \ldots, f_N$, to get

(5.2) $$f_0 = G_N + H_N f_{N+1} \qquad (N \geq 0),$$

where

(5.3) $$G_0 = L_0, \qquad H_0 = M_0.$$

By (5.2) and (5.1) (with $n = N+1$), we have

$$f_0 = G_N + H_N(L_{N+1} + M_{N+1}f_{N+2}) = (G_N + H_N L_{N+1}) + (M_{N+1}H_N)f_{N+2}$$
$$= G_{N+1} + H_{N+1}f_{N+2}.$$

Hence

(5.4) $$G_{N+1} = G_N + L_{N+1}H_N,$$

(5.5)
$$H_{N+1} = M_{N+1}H_N.$$

From (5.5) and (5.3) we find

(5.6)
$$H_N = M_0 M_1 M_2 \cdots M_N,$$

and summing (5.4),

(5.7)
$$G_N = L_0 + \sum_{r=0}^{N-1} L_{r+1}(M_0 M_1 \cdots M_r).$$

Now suppose that $f_N \to f$, $G_N \to G$, $H_N \to H$. Then

(5.8)
$$f_0 = G + Hf.$$

It will sometimes happen that f_N, G_N, H_N do not converge separately; it may then be possible to throw the combination (5.2) into such a form that its limit (which obviously exists) is readily recognizable, or to allow N to increase over some particular set of integers.

6. Application of iteration $(t \to tq)$. Let $f_0 = F(a, b; t)$, $f_n = F(a, b; tq^n)$, $L_n = (1-b)/(1-tq^n)$, $M_n = (b - atq^{n+1})/(1-tq^n)$. Replacing t by tq^n in (2.4), we obtain (5.1). We find that $L_n \to 1 - b$, $M_n \to b$, $f_n \to F(a, b; 0) = 1$, and

$$H_n = \prod_{r=0}^{n} \frac{(b - atq^{r+1})}{(1 - tq^r)}.$$

We must now consider several cases, depending on the value of b.

First, suppose that $b = 0$. Then $H_n \to 0$, so $F(a, 0; t) = \lim_{n \to \infty} G_n$; that is,

(6.1)
$$F(a, 0; t) = \frac{1}{1-t} \sum_{n \geq 0} \frac{(-at)^n q^{(n^2+n)/2}}{(tq)_n}.$$

Next, suppose that $b = 1$. Then $L_n = 0$, $G_n = 0$, and $F(a, 1; t) = \lim_{n \to \infty} H_n$; that is,

(6.2)
$$F(a, 1; t) = \frac{(atq)_\infty}{(t)_\infty}.$$

Equation (6.2) contains as special cases a number of well-known identities, three of which we state for later reference. The case $a = 0$ yields

(6.21)
$$F(0, 1; t) = \sum_{n \geq 0} \frac{t^n}{(q)_n} = \frac{1}{(t)_\infty}.$$

Now put $a = q^m$, $m \geq 0$. Then, recalling (1.3), we have

(6.22)
$$F(q^m, 1; t) = \sum_{n \geq 0} \begin{bmatrix} m + n \\ n \end{bmatrix} t^n = \frac{1}{(t)_{m+1}} \qquad (m \geq 0),$$

or by a slight change,

(6.221)
$$\frac{1}{(uq)_N} = \sum_{n \geq 0} \begin{bmatrix} N + n - 1 \\ n \end{bmatrix} (uq)^n \qquad (N \geq 1).$$

Finally, putting $a = q^{-(N+1)}$, $N \geq 0$, replacing q by q^{-1}, and simplifying, we get

$$(6.23) \qquad (tq)_N = \sum_{k=0}^{N} \begin{bmatrix} N \\ k \end{bmatrix} (-t)^k q^{(k^2+k)/2} \qquad (N \geq 0).$$

Equations (6.221) and (6.23) are the basic analogues of the binomial series, to which they reduce termwise for $q = 1$. The condition for convergence in (6.22) is $|t| < 1$, and in (6.221), $|u| < |q|^{-1}$. The series in (6.1) converges for all a and t, $t \neq q^{-m}$, $m \geq 0$.

Returning to our discussion of cases, suppose that $0 < |b| < 1$. Then $H_n \to 0$, and factoring b^n from the nth term in the resulting series yields

$$(6.3) \qquad F(a, b; t) = \frac{1-b}{1-t} F\left(\frac{at}{b}, t; b \right).$$

Thus the function $(1 - t)F(a, b; t)$ is invariant under the involution $a' = at/b$, $b' = t$, $t' = b$. Equation (6.3), proved for $0 < |b| < 1$, holds for all values for which both sides are regular.

If we multiply both sides of (6.3) by $(1 - t)$ and let $t \to 1$, we get

$$(6.31) \qquad \lim_{t \to 1} (1 - t)F(a, b; t) = (1 - b)F(a/b, 1; b) = \frac{(aq)_\infty}{(bq)_\infty}$$

by virtue of (6.2). This could have been proved directly, of course.

7. Iteration of $b \to bq$. Here we assume $at \neq 0$, and take $f_n = F(a, bq^n; t)$,

$$L_n = \frac{bq^n}{bq^n - at}, \qquad M_n = \frac{(bq^n - a)t}{(1 - bq^{n+1})(bq^n - at)}.$$

Then equation (4.4), with b replaced by bq^n, assumes the form (5.1). We find

$$H_n = \prod_{r=0}^{n} \frac{(1 - (b/a)q^r)}{(1 - bq^{r+1})(1 - (b/at)q^r)},$$

$$G_n = -\frac{(b/at)}{1 - (b/at)} \sum_{k=0}^{n} \frac{(b/a)_k}{(bq)_k (bq/at)_k} q^k.$$

Also $f_n \to F(a, 0; t)$. If, by a slight change in notation, we set

$$(7.1) \qquad S = F(a, 0; t), \qquad H = \frac{(b/a)_\infty}{(bq)_\infty (b/at)_\infty},$$

$$G = \frac{(b/at)}{1 - (b/at)} \sum_{n \geq 0} \frac{(b/a)_n}{(bq)_n (bq/at)_n} q^n, \qquad F = F(a, b; t),$$

then we have

$$(7.2) \qquad F + G = HS.$$

Two special cases are of interest here. First, multiply (7.2) by $1 - (b/at)$ and set $a = b/t$ to get

$$(7.3) \qquad \sum_{n \geq 0} \frac{(t)_n}{(bq)_n (q)_n} q^n = \frac{(t)_\infty}{(bq)_\infty (q)_\infty} F(b/t, 0; t).$$

Recalling from (6.1) that

$$(7.31) \qquad F(b/t, 0; t) = \frac{1}{1-t} \sum_{n \geq 0} \frac{(-b)^n}{(tq)_n} q^{(n^2+n)/2},$$

and letting $t \to 0$ in (7.3) and (7.31), we obtain

$$(7.32) \qquad \sum_{n \geq 0} \frac{q^n}{(bq)_n (q)_n} = \frac{1}{(bq)_\infty (q)_\infty} \sum_{n \geq 0} (-b)^n q^{(n^2+n/2)}.$$

For $b = -1$, the left side becomes $F(0, 1; q : q^2)$, which can be expressed as a product by (6.21). Simplifying yields

$$(7.321) \qquad \sum_{n \geq 0} q^{(n^2+n)/2} = \prod_{n \geq 1} \frac{(1-q^{2n})}{(1-q^{2n-1})},$$

a well-known formula due to Gauss. For $b = 1$, we have

$$(7.322) \qquad \sum_{n \geq 0} \frac{q^n}{(q)_n^2} = \frac{1}{(q)_\infty^2} \sum_{n \geq 0} (-1)^n q^{(n^2+n)/2}.$$

This result appears to be new. It is related to a similar formula due to Auluck [4]. If we put $b = q^{-1/2}$ in (7.32) and then replace q by q^2, we get

$$(7.323) \qquad \sum_{n \geq 0} \frac{q^{2n}}{(q)_{2n}} = \frac{1}{(q)_\infty} \sum_{n \geq 0} (-1)^n q^{n^2}.$$

This has an interesting arithmetic interpretation, which we shall discuss later (22.21). It is easy to see that the left side of (7.323) can be written as

$$\frac{1}{2} \{ F(0, 1; q) + F(0, 1; -q) \} = \frac{1}{2} \left\{ \frac{1}{(q)_\infty} + \frac{1}{(-q)_\infty} \right\}.$$

Thus, eventually, we get

$$(7.324) \qquad \sum_{-\infty}^{+\infty} (-1)^n q^{n^2} = \prod_{n \geq 1} \left(\frac{1-q^n}{1+q^n} \right) = \prod_{n \geq 1} (1-q^{2n})(1-q^{2n-1})^2,$$

or, on replacing q by $-q$,

$$(7.325) \qquad \sum_{-\infty}^{+\infty} q^{n^2} = \prod_{n \geq 1} (1-q^{2n})(1+q^{2n-1})^2.$$

These last two are standard results in the theory of the elliptic theta-functions, and are special cases of an identity of Jacobi's that we shall prove later (§17).

The second special case of (7.2) is obtained by setting $a = b^2/t$; then

$$S = F(b^2/t, 0; t) = \sum_{n \geq 0} \frac{(-1)^n b^{2n} q^{(n^2+n)/2}}{(t)_{n+1}},$$

$$(7.4) \qquad \begin{aligned} H &= \frac{(t/b)_\infty}{(bq)_\infty (b^{-1})_\infty}, \\ G &= \frac{b^{-1}}{1-b^{-1}} \sum_{n \geq 0} \frac{(t/b)_n}{(bq)_n (b^{-1}q)_n} q^n, \\ F &= F(b^2/t, b; t) = \left(\frac{1-b}{1-t} \right) F(b, t; b), \end{aligned}$$

the second form of F being derived from (6.3). Now $t \to 0$ leads to

(7.5)
$$S = \sum_{n \geq 0} (-1)^n b^{2n} q^{(n^2+n)/2},$$

$$H = (qb)_\infty^{-1} (b^{-1})_\infty^{-1},$$

$$G = -\frac{1}{(1-b)} \sum_{n \geq 0} \frac{q^n}{(bq)_n (b^{-1}q)_n},$$

$$F = (1-b)F(b,0;b).$$

It is possible to derive a simple power series expansion for $h(b) \equiv (1-b)F(b,0;b)$. Indeed, (4.6) yields

(7.6)
$$h(b) = 1 - b^2 q - b^3 q^2 h(bq).$$

If we substitute a power series for $h(b)$ in (7.6) and equate coefficients, we obtain

(7.7) $\quad (1-b)F(b,0;b) = \sum_{n \geq 0} (-1)^n q^{(3n^2+n)/2} b^{3n} + \sum_{n \geq 1} (-1)^n q^{(3n^2-n)/2} b^{3n-1}.$

An important limiting case is $b \to 1$. Using (6.31), we get

(7.8)
$$\prod_{n \geq 1} (1-q^n) = \sum_{-\infty}^{+\infty} (-1)^n q^{(3n^2+n)/2},$$

a famous identity due to Euler.

8. Consequences of §7. In (7.1) and (7.2), set $a = -b/ut$ to obtain

(8.1)
$$S = F(-b/ut, 0; t) = \frac{1}{1-t} \sum_{n \geq 0} \frac{(bu^{-1})^n}{(tq)_n} q^{(n^2+n)/2},$$

$$H = \frac{(ut)_\infty}{(bq)_\infty (u)_\infty},$$

$$G = -\frac{u}{1+u} \sum_{n \geq 0} \frac{(ut)_n}{(bq)_n (-uq)_n} q^n,$$

$$F = F(-b/ut, b; t) = \left(\frac{1-b}{1-t}\right) F(-u^{-1}, t; b),$$

with $F + G = HS$; the second forms of S and F are applications of (6.1) and (6.3) respectively. Now let $t \to 0$; we get

(8.2)
$$S = \sum_{n \geq 0} (bu^{-1})^n q^{(n^2+n)/2},$$

$$H = ((bq)_\infty (u)_\infty)^{-1},$$

$$G = -\frac{u}{1+u} \sum_{n \geq 0} \frac{q^n}{(bq)_n (-uq)_n},$$

$$F = (1-b)F(-u^{-1}, 0; b).$$

For $u = b$, (8.2) simplifies to

$$S = \sum_{n \geq 0} q^{(n^2+n)/2},$$

(8.3)

$$H = \frac{1}{1+b} \prod_{n \geq 1} \frac{1}{1 - b^2 q^{2n}},$$

$$G = -\frac{b}{1+b} F(0, b^2; q : q^2),$$

$$F = (1-b)F(-b^{-1}, 0; b) = \sum_{n \geq 0} \frac{q^{(n^2+n)/2}}{(bq)_n}.$$

The equation $F + G = HS$ now reduces to

(8.4)

$$F(-b^{-1}, 0; b) = \frac{1}{1-b} \sum_{n \geq 0} \frac{q^{(n^2+n)/2}}{(bq)_n}$$

$$= S \prod_{n \geq 0} \frac{1}{1 - b^2 q^{2n}} + \frac{b}{1-b^2} F(0, b^2; q : q^2).$$

If we expand the second member of (8.4) in powers of b by (6.221), we get

(8.41)

$$F(-b^{-1}, 0; b) = \sum_{k \geq 0} V_k b^k,$$

where

(8.42)

$$V_k = \sum_{n \geq 0} \begin{bmatrix} n+k \\ n \end{bmatrix} q^{(n^2+n)/2}.$$

On expanding the third member of (8.4) by (6.21), (6.3), and (1.1), we obtain

(8.43)

$$F(-b^{-1}, 0; b) = S \sum_{n \geq 0} \frac{b^{2n}}{(q^2 : q^2)_n} + \sum_{n \geq 0} \frac{b^{2n+1}}{(1-q)(1-q^3)(1-q^5) \cdots (1-q^{2n+1})}.$$

Equating coefficients of b^k in (8.41) and (8.43) yields

(8.5a)

$$V_{2k} = \frac{S}{(q^2 : q^2)_k},$$

(8.5b)

$$V_{2k+1} = \frac{1}{(1-q)(1-q^3) \cdots (1-q^{2k+1})}.$$

It is easy to see that V_k converges as $k \to \infty$; this yields another proof of Gauss's formula (7.321).

The results of this section will be used later to derive a theorem on partitions, stated in [16].

9. Further consequences of §7. In (7.1) and (7.2), replace b by aq to obtain

$$(9.1) \quad tF(a, 0; t) = \frac{(aq)_\infty (t^{-1}q)_\infty}{(q)_\infty} \left\{ \frac{t}{1-aq} F(a, aq; t) + \sum_{n \geq 1} \frac{(q)_{n-1}}{(aq)_n (t^{-1}q)_n} q^n \right\}.$$

Now put $a = -t^{-1}$:

$$tF(-t^{-1}, 0; t) = \frac{(-t^{-1}q)_\infty (t^{-1}q)_\infty}{(q)_\infty} \left\{ \frac{t}{1 + t^{-1}q} F(-t^{-1}, -t^{-1}q; t) \right.$$

$$\left. + \sum_{n \geq 1} \frac{(q)_{n-1}}{(-t^{-1}q)_n (t^{-1}q)_n} q^n \right\}.$$

The infinite product on the right is obviously an even function of t, and so is the series. Hence

$$(9.12) \quad t\{F(-t^{-1}, 0; t) + F(t^{-1}, 0; -t)\}$$

$$= \frac{(t^{-2}q^2 : q^2)_\infty}{(q)_\infty} \left\{ \frac{t}{1 + t^{-1}q} F(-t^{-1}, -t^{-1}q; t) - \frac{(-t)}{1 - t^{-1}q} F(t^{-1}, t^{-1}q; -t) \right\}.$$

Now by (6.3),

$$\frac{t}{1 + t^{-1}q} F(-t^{-1}, -t^{-1}q; t) = \frac{t}{1 - t} F(tq^{-1}, t; -t^{-1}q)$$

$$= \frac{t}{1 - t} + t \sum_{n \geq 1} \frac{(-t^{-1}q)^n}{1 - tq^n},$$

the last expansion being valid for $|t| > |q|$. If we subtract the similar expression with t replaced by $-t$, we get

$$\frac{2t}{1 - t^2} + t \sum_{n \geq 1} t^{-n} q^n \left\{ \frac{(-1)^n}{1 - tq^n} + \frac{1}{1 + tq^n} \right\}.$$

Hence, in the ring $|q| < |t| < |q|^{-1}$, we can expand in a double series

$$\frac{2t}{1 - t^2} + \sum_{k,n \geq 1} q^{kn} t^{k-n} \{(-1)^n - (-1)^k\}$$

$$= \frac{2t}{1 - t^2} + \sum_{N \geq 1} q^N \sum_{kn = N} t^{k-n} \{(-1)^n - (-1)^k\}$$

$$= \frac{2t}{1 - t^2} + \sum_{N \geq 1} q^{2N} \sum_{kn = 2N} t^{k-n} \{(-1)^n - (-1)^k\},$$

since the inner sum vanishes unless n and k are of opposite parity, so that their product must be even. This inner sum may be written

$$2 \sum_{\substack{kn = 2N \\ k \text{ odd}}} t^{k-n} - 2 \sum_{\substack{kn = 2N \\ n \text{ odd}}} t^{k-n}.$$

Letting ω denote an odd divisor of $2N$, therefore of N, we get

$$-2 \sum_{\omega | N} (t^{2N/\omega - \omega} - t^{-2N/\omega + \omega}).$$

Hence

$$(9.13) \qquad t\{F(-t^{-1},0;t) + F(t^{-1},0;-t)\}$$

$$= \frac{(t^{-2}q^2;q^2)_\infty}{(q)_\infty}\left\{\frac{2t}{1-t^2} - 2\sum_{N\geq 1} q^{2N}\sum_{\omega|N}(t^{2N/\omega-\omega} - t^{-2N/\omega+\omega})\right\}.$$

The quantity in brackets on the left side of (9.13) is twice the even part of $F(-t^{-1},0;t)$; by (8.4) this is $2S(1-t^2)^{-1}(t^2q^2;q^2)_\infty^{-1}$. Recalling the value of S from (8.3) and (7.321), we have $S(q)_\infty = (q^2;q^2)_\infty^2$; finally,

$$(9.14) \qquad \left(\frac{2t}{1-t^2}\right)\frac{(q^2;q^2)_\infty^2}{(t^{-2}q^2;q^2)_\infty(t^2q^2;q^2)_\infty}$$

$$= \frac{2t}{1-t^2} - 2\sum_{N\geq 1}q^{2N}\sum_{\omega|N}(t^{2N/\omega-\omega} - t^{-2N/\omega+\omega}),$$

the formula being valid in the ring $|q| < |t| < |q|^{-1}$. Replacing q^2 by q throughout, we have, for $|q| < |t|^2 < |q|^{-1}$,

$$(9.2) \qquad \frac{(q)_\infty^2}{(t^{-2}q)_\infty(t^2q)_\infty} = 1 + (t-t^{-1})\sum_{N\geq 1}q^N\sum_{\omega|N}(t^{2N/\omega-\omega} - t^{-2N/\omega+\omega}).$$

If we let $t = e^{iu}$, u real, this becomes

$$(9.3) \qquad \frac{(q)_\infty^2}{\prod_{n\geq 1}(1 - 2q^n\cos 2u + q^{2n})} = 1 - 4\sin u\sum_{N\geq 1}q^N\sum_{\omega|N}\sin\left(\frac{2N}{\omega} - \omega\right)u.$$

This formula contains a wealth of information, some of which we shall unearth later. It is essentially the Fourier expansion of $\vartheta_1'(0\mid\tau)/\vartheta_1(u\mid\tau)$, with $q = \exp(2\pi i\tau)$.

10. A product-series identity. We shall now prove an identity that connects certain infinite products with series whose coefficients have arithmetic significance.

Let p be an arbitrary integer greater than 1, and let r be chosen relatively prime to $2p$, with $0 < r < p$. From (9.12)–(9.13), it is easy to see that

$$(10.1) \qquad \frac{2t}{1-t^2}\prod_{n\geq 1}\frac{(1-q^{2n})^2}{(1-t^2q^{2n})(1-t^{-2}q^{2n})}$$

$$= \frac{2t}{1-t^2} + \sum_{k,n\geq 1}q^{kn}t^{k-n}((-1)^n - (-1)^k),$$

for $|q| < |t| < |q|^{-1}$. Replace q by $q^{p/2}$, t by $q^{r/2}$, to get

(10.2)

$$2q^{r/2} \prod_{n\geq 1} \frac{(1-q^{pn})^2}{(1-q^{pn-r})(1-q^{pn-p+r})}$$

$$= \frac{2q^{r/2}}{1-q^r} + \sum_{k,n\geq 1} q^{(pkn+r(k-n))/2}((-1)^n - (-1)^k)$$

$$= \frac{2q^{r/2}}{1-q^r} + 2\sum_{\substack{n \text{ even} \\ k \text{ odd}}} q^{(pkn+r(k-n))/2} - 2\sum_{\substack{n \text{ odd} \\ k \text{ even}}} q^{(pkn+r(k-n))/2}$$

$$= \frac{2q^{r/2}}{1-q^r} + 2\sum_{\substack{n \text{ even} \\ k \text{ odd}}} \{q^{(pkn+r(k-n))/2} - q^{(pkn-r(k-n))/2}\}$$

$$= \frac{2q^{r/2}}{1-q^r} + 2\sum_{n,k\geq 1} \{q^{pn(2k-1)+r(k-n)-r/2} - q^{pn(2k-1)-r(k-n)+r/2}\}.$$

Dividing through by $2q^{r/2}$, we have

$$\prod_{n\geq 1} \frac{(1-q^{pn})^2}{(1-q^{pn-r})(1-q^{pn-p+r})}$$

(10.21)

$$= \frac{1}{1-q^r} + \sum_{n,k\geq 1} \{q^{2pkn-pn-rn+(k-1)r} - q^{2pkn-pn+rn-kr}\}$$

$$= \frac{1}{1-q^r} + \Phi_1(q) - \Phi_2(q),$$

say. It is easily verified that

(10.3) $$\Phi_1(q^{2p}) = q^{r^2-rp} \sum_{n,k\geq 1} q^{(2pn+r)(2pk-p-r)} = q^{r^2-rp} {\sum}' q^{MN},$$

where $M = 2pn + r$ $(n \geq 1)$, $N = 2p(k-1) + p - r$ $(k \geq 1)$. That is, $M \equiv r$ (mod $2p$), $M > r$; $N \equiv p - r$ (mod $2p$), $N > 0$. Taking into account the term $(1-q^r)^{-1}$ in (10.21), we get

(10.4)

$$\frac{1}{1-q^{2pr}} + \Phi_1(q^{2p}) = q^{r^2-rp} \left\{ \sum_{k\geq 0} q^{2prk+rp-r^2} + {\sum}' q^{MN} \right\}$$

$$= q^{r^2-rp} \sum q^{MN},$$

where now the sum runs over all $M, N > 0$ satisfying $M \equiv r$, $N \equiv p-r$ (mod $2p$). Collecting all the terms which involve q^{2pn}, $2pn = MN - r(p-r)$, $n \geq 0$, we see that M runs over all the positive divisors of $2pn + r(p-r)$ for which the complementary divisor is congruent to $p - r$, and $M \equiv r$ (mod $2p$). The latter condition automatically implies the former, since $(r, 2p) = 1$. Hence

(10.5) $$\frac{1}{1-q^{2pr}} + \Phi_1(q^{2p}) = \sum_{n\geq 0} q^{2pn} \sum_{\substack{d|2pn+r(p-r) \\ d\equiv r \ (\text{mod } 2p)}} 1.$$

Similarly, it is easy to show that

$$(10.51) \qquad \Phi_2(q^{2p}) = \sum_{n \geq 0} q^{2pn} \sum_{\substack{d \mid 2pn + r(p-r) \\ d \equiv -r \ (\text{mod } 2p)}} 1.$$

Finally, replacing q^{2p} by q, we have, for $0 < r < p$, $(r, 2p) = 1$,

$$(10.6) \qquad \prod_{n \geq 1} \frac{(1 - q^{pn})^2}{(1 - q^{pn-r})(1 - q^{pn-p+r})} = \sum_{n \geq 0} E_r(2pn + r(p-r); 2p)q^n,$$

where we define

$$(10.7) \qquad E_r(N; m) = \sum_{\substack{d \mid N \\ d \equiv r \ (\text{mod } m)}} 1 - \sum_{\substack{d \mid N \\ d \equiv -r \ (\text{mod } m)}} 1;$$

in general,

$$(10.71) \qquad E_{r,s,\ldots,z}(N; m) = E_r(N; m) + E_s(N; m) + \cdots + E_z(N; m).$$

Formula (10.6) is the one we set out to obtain. It is clear that it should be capable of many uses. Two types immediately present themselves: (a) those cases in which the product has another expansion, so that we are led to an arithmetic identity by equating coefficients, and (b) those cases in which the coefficients E_r lend themselves to further transformations, which are then reflected in identities among infinite products. We believe that (10.6) is new.

11. Iteration of $a \to aq$. A routine application of iteration to equation (4.5) yields

$$(11.1) \qquad F + G = HS,$$

where

$$(11.2) \qquad \begin{aligned} S &= F(0, b; t), \\ H &= \frac{(aq)_\infty (atq/b)_\infty}{(aq/b)_\infty}, \\ F &= F(a, b; t), \\ G &= \frac{(1 - b)(aq/b)}{(1 - aq/b)} \sum_{n \geq 0} \frac{(aq)_n (atq/b)_n}{(atq^2/b)_n} q^n. \end{aligned}$$

For $a = b$, this reduces to

$$(11.3) \qquad \frac{1}{1 - t} \sum_{n \geq 0} \frac{(b)_n (t)_n}{(q)_n} q^n = \frac{(bq)_\infty (tq)_\infty}{(q)_\infty} F(0, b; t).$$

An interesting special case is obtained if we set $b = t^{-1}$ in (11.3):

$$(11.4) \qquad \frac{(t)_\infty (t^{-1}q)_\infty}{(q)_\infty} F(0, t^{-1}; t) = \sum_{n \geq 0} \frac{(t)_n (t^{-1})_n}{(q)_n} q^n.$$

12. Iteration of (b, t) $\rightarrow (bq, tq)$. If we write

$$R(a, b; t) = (1 - t)F(a, b; t),$$

equation (4.3) becomes

(12.1) $$R(a, b; t) = 1 + \frac{(b - a)tq}{(1 - bq)(1 - tq)}R(a, bq; tq).$$

Carrying through the iteration method, and observing that $H_n \rightarrow 0$, $f_n = R(a, bq^n; tq^n) \rightarrow 1$, we obtain

(12.2) $$(1 - t)F(a, b; t) = \sum_{n \geq 0} \frac{(b/a)_n}{(bq)_n (tq)_n}(-at)^n q^{(n^2+n)/2}.$$

This is a generalization of (6.1), to which it reduces for $b = 0$. If we let $a \rightarrow 0$, however, we obtain

(12.3) $$(1 - t)F(0, b; t) = \sum_{n \geq 0} \frac{(bt)^n q^{n^2}}{(bq)_n (tq)_n}.$$

Now $t \rightarrow 1$ yields, in view of (6.32),

(12.31) $$\frac{1}{(bq)_\infty} = \sum_{n \geq 0} \frac{b^n q^{n^2}}{(bq)_n (q)_n}.$$

For $b = 1$ we get another well-known identity:

(12.311) $$\frac{1}{(q)_\infty} = \sum_{n \geq 0} \frac{q^{n^2}}{(q)_n^2}.$$

If we let $b = t^{-1} = e^{-i\theta}$, (12.3) reduces to

(12.32) $$\begin{aligned} &(1 - e^{i\theta})F(0, e^{-i\theta}; e^{i\theta}) \\ &= 1 + \sum_{n \geq 1} \frac{q^{n^2}}{(1 - 2q\cos\theta + q^2)\cdots(1 - 2q^n\cos\theta + q^{2n})}. \end{aligned}$$

Put $b = t$ in (12.3). Then

(12.33) $$(1 - t)F(0, t; t) = \sum_{n \geq 0} \frac{t^{2n} q^{n^2}}{(tq)_n^2}.$$

Let r and m be positive integers, and replace q by q^m, t by q^r in (12.33). Then, using definition (1.1) for the left side, we obtain the curious identity

(12.331)
$$\sum_{n \geq 0} \frac{q^{rn}}{(1 - q^r)(1 - q^{m+r})\cdots(1 - q^{nm+r})}$$
$$= \sum_{n \geq 0} \frac{q^{mn^2+2rn}}{\{(1 - q^r)(1 - q^{m+r})\cdots(1 - q^{nm+r})\}^2},$$

of which (12.311) is a special case.

Let us define

$$(12.4) \qquad \Phi(q) = \frac{q}{1-q} F(1, q; q) = \sum_{n \geq 1} \frac{q^n}{1 - q^n}.$$

It is well-known and easy to prove that $\Phi(q)$ is the generating function for $d(N)$, the number of divisors of N:

$$(12.41) \qquad \Phi(q) = \sum_{N \geq 1} d(N) q^N.$$

Putting $a = 1$, $b = t = q$ in (12.2), we find

$$(12.42) \qquad \Phi(q) = \sum_{n \geq 1} (-1)^{n-1} \frac{q^{(n^2+n)/2}}{(q)_n (1 - q^n)}.$$

We may expand the extraneous factor $(1 - q^n)^{-1}$ to get the double series

$$(12.43) \qquad \Phi(q) = \sum_{k \geq 0} \sum_{n \geq 1} (-1)^{n-1} \frac{q^{kn} q^{(n^2+n)/2}}{(q)_n}.$$

To evaluate the inner sum, let $t \to 1$ in (6.1) and use (6.31) to derive the useful formula

$$(12.44) \qquad (aq)_\infty = \sum_{n \geq 0} \frac{(-a)^n q^{(n^2+n)/2}}{(q)_n}.$$

Replacing a by q^k reduces (12.43) to the form

$$(12.45) \qquad \Phi(q) = \sum_{k \geq 0} (1 - (q^{k+1})_\infty) = \sum_{k \geq 0} \left(1 - \prod_{n > k} (1 - q^n) \right).$$

13. Iteration of $(a, t) \to (aq, tq)$. For $0 < |b| < 1$, the iteration method applied to (4.6) yields

$$(13.1) \qquad \left(\frac{1-t}{1-b} \right) F(a, b; t) = \sum_{n \geq 0} \frac{(aq)_n (atq/b)_{2n}}{(tq)_n (aq/b)_n} b^n$$

$$- aq \sum_{n \geq 0} \frac{(aq)_n (atq/b)_{2n+1}}{(tq)_n (aq/b)_{n+1}} (bq)^n.$$

By (6.3), the left side of (13.1) is $F(at/b, t; b)$; making the substitution $a' = at/b$, $b' = t$, $t' = b$, then dropping the primes, we obtain a new form:

$$(13.2) \qquad F(a, b; t) = \sum_{n \geq 0} \frac{(atq/b)_n (aq)_{2n}}{(bq)_n (aq/b)_n} t^n$$

$$- \frac{atq}{b} \sum_{n \geq 0} \frac{(atq/b)_n (aq)_{2n+1}}{(bq)_n (aq/b)_{n+1}} (tq)^n,$$

valid for $|t| < 1$. Letting $b \to 0$ in (13.1), we find

$$(13.3) \qquad (1 - t) F(a, 0; t) = \sum_{n > 0} \frac{(aq)_n}{(tq)_n} (-at^2)^n (1 - atq^{2n+1}) q^{(3n^2+n)/2}.$$

If we now let $t = a$, we recover (7.7), and if we let $t \rightarrow 1$, we get

$$(13.4) \qquad (aq)_\infty = \sum_{n \geq 0} \frac{(aq)_n}{(q)_n} (-a)^n (1 - aq^{2n+1}) q^{(3n^2+n)/2}.$$

On the other hand, if we let $t \rightarrow 1$ in (13.1), we get the more general

$$(13.5) \qquad \begin{aligned} \frac{(aq)_\infty}{(b)_\infty} &= \sum_{n \geq 0} \frac{(aq)_n (aq/b)_{2n}}{(q)_n (aq/b)_n} b^n \\ &\quad - aq \sum_{n \geq 0} \frac{(aq)_n (aq/b)_{2n+1}}{(q)_n (aq/b)_{n+1}} (bq)^n. \end{aligned}$$

For $|a| < 1$ and $b = a$, (13.5) becomes

$$(13.51) \qquad (1-a)^{-1} = \sum_{n \geq 0} \begin{bmatrix} 2n \\ n \end{bmatrix} (aq)_n a^n - \sum_{n \geq 0} \begin{bmatrix} 2n+1 \\ n \end{bmatrix} (aq)_n (aq)^{n+1}.$$

As a check, if we let $q = 1$, here, as we may, we obtain the not completely trivial identity

$$(13.52) \qquad (1-a)^{-1} = \sum_{n \geq 0} \binom{2n}{n} z^n - a \sum_{n \geq 0} \binom{2n+1}{n} z^n,$$

where we have set $z = a(1 - a)$.

14. Iteration of $(a, b, t) \rightarrow (aq, bq, tq)$. If we assume that $b \neq 0$, iteration of (4.1) is easily seen to yield

$$(14.1) \qquad (1-t)F(a, b; t) = \sum_{n \geq 0} \frac{(aq)_n (atq/b)_n}{(bq)_n (tq)_n} (1 - atq^{2n+1})(bt)^n q^{n^2}.$$

This identity contains much information about the function $F(a, b; t)$, and many of our earlier results are easily derived from it. For example, (6.3) is an immediate consequence of the fact that the right side is formally invariant under the transformation $(a, b, t) \rightarrow (at/b, t, b)$. If we let $t \rightarrow 1$ in (14.1) and use (6.31), we get

$$(14.2) \qquad \frac{(aq)_\infty}{(bq)_\infty} = \sum_{n \geq 0} \frac{(aq)_n (aq/b)_n}{(bq)_n (q)_n} (1 - aq^{2n+1}) b^n q^{n^2},$$

of which (13.4) is a limiting case, for $b \rightarrow 0$. Equation (12.3) also follows directly from $a = 0$ in (14.1).

Set $t = a$ in (14.1):

$$(14.3) \qquad (1-a)F(a, b; a) = \sum_{n \geq 0} \frac{(a^2 q/b)_n}{(bq)_n} (1 - a^2 q^{2n+1})(ab)^n q^{n^2}.$$

This simplifies if $a^2/b = b$, that is, if $b = \pm a$. For $b = -a$, we have

$$(14.31) \qquad (1-a)F(a, -a; a) = 1 + 2 \sum_{n \geq 1} (-a^2)^n q^{n^2}.$$

Thus, $a \to 1$ recovers for us equation (7.324); $a = i$ yields

$$(14.32) \qquad (1-i)F(i,-i;i) = \sum_{-\infty}^{+\infty} q^{n^2}.$$

In (14.1), replace q by q^2 and set $a = bq^{-1}$, $t = bq$; we find easily that

$$(14.4) \qquad F(bq^{-1}, b; bq : q^2) = \sum_{n \geq 0} b^n q^{(n^2+n)/2}.$$

Here we can recover Gauss's formula (7.321) by setting $b = 1$ and using (6.2). Another curious result is

$$(14.41) \qquad F(q^{-3}, q^{-1}; q : q^4) = \sum_{n \geq 0} q^{n^2},$$

that is,

$$(14.42) \qquad 1 + \frac{(1-q)}{(1-q^3)}q + \frac{(1-q)(1-q^5)}{(1-q^3)(1-q^7)}q^2 + \cdots = 1 + q + q^4 + q^9 + \cdots.$$

An identity of similar nature is obtained by combining (12.2) for $a = -b = t = i$ with (14.32):

$$(14.43) \qquad \begin{aligned} & 1 + \frac{q}{1+q^2} + \frac{(1+q)q^3}{(1+q^2)(1+q^4)} \\ & + \frac{(1+q)(1+q^2)q^6}{(1+q^2)(1+q^4)(1+q^6)} + \cdots = \sum_{n \geq 0} q^{n^2}. \end{aligned}$$

We close this section with one more example of the versatility of (14.1). Put $t = b$ there:

$$(14.5) \qquad (1-b)F(a,b;b) = \sum_{n \geq 0} \left(\frac{(aq)_n}{(bq)_n} \right)^2 (1 - abq^{2n+1})b^{2n}q^{n^2}.$$

Now if we set $a = 1$, $b = q$ and use (12.4) and (12.41), we obtain the well-known result of Clausen [10]

$$(14.51) \qquad \sum_{N \geq 1} d(N)q^N = \sum_{n \geq 1} \left(\frac{1+q^n}{1-q^n} \right) q^{n^2},$$

which can also be proved directly by expanding the right side in powers of q.

15. A special development. The function

$$(15.1) \qquad G(t) \equiv (t)_\infty F(a,b;t) \equiv \sum_{n \geq 0} A_n t^n,$$

with A_n independent of t, is clearly entire. We shall determine the coefficients A_n. By using (1.1) and (12.44), we see directly that

$$(15.2) \qquad A_N = \sum_{k=0}^{N} (-1)^k \frac{(aq)_{N-k}}{(bq)_{N-k}} \frac{q^{(k^2-k)/2}}{(q)_k}.$$

On the other hand, substitution into (2.4) shows that

(15.3) $$G(t) = (1 - b)(tq)_\infty + (b - atq)G(tq),$$

from which we deduce, by equating coefficients and using (12.44) again, that for $n \geq 1$,

(15.31) $$A_n(1 - bq^n) = -aq^n A_{n-1} + (-1)^n(1 - b)\frac{q^{(n^2+n)/2}}{(q)_n}.$$

The obvious substitution

$$A_n = \frac{(-a)^n q^{(n^2+n)/2}}{(bq)_n} B_n$$

reduces (15.31) to

$$B_n = B_{n-1} + \frac{(b)_n}{(q)_n} a^{-n},$$

whence, summing for $n = 1, 2, \ldots, N$, we obtain

(15.4) $$B_N = \sum_{n=0}^{N} \frac{(b)_n}{(q)_n} a^{-n},$$

(15.41) $$A_N = \frac{(-a)^N q^{(N^2+N)/2}}{(bq)_N} \sum_{n=0}^{N} \frac{(b)_n}{(q)_n} a^{-n}.$$

Thus we have the identity

(15.5) $$(t)_\infty F(a, b; t) = \sum_{N \geq 0} \frac{(-at)^N q^{(N^2+N)/2}}{(bq)_N} \sum_{n=0}^{N} \frac{(b)_n}{(q)_n} a^{-n}.$$

Two special cases are of interest here:

(15.51) $$(t)_\infty F(0, b; t) = (1 - b) \sum_{N \geq 0} \frac{(-t)^N q^{(N^2+N)/2}}{(1 - bq^N)(q)_N},$$

(15.52) $$(t)_\infty F(a, 0; t) = \sum_{N \geq 0} (-t)^N q^{(N^2+N)/2} \sum_{n=0}^{N} \frac{a^{N-n}}{(q)_n}.$$

16. The partial-fraction decomposition. Define

(16.1) $$C_n = \lim_{t \to q^{-n}} (1 - tq^n)F(a, b; t) \qquad (n \geq 0).$$

From (2.4) it is easy to see that

$$(1 - q^{-n})C_n = (b - aq^{1-n})C_{n-1} \qquad (n \geq 1),$$

so that

$$C_n = C_0 \frac{(b/a)_n}{(q)_n}(aq)^n,$$

and from (6.31)

(16.2) $$C_n = \frac{(aq)_\infty}{(bq)_\infty} \cdot \frac{(b/a)_n}{(q)_n}(aq)^n \qquad (n \geq 0).$$

Equations (16.1) and (16.2) suggest that

$$(16.3) \qquad F(a,b;t) = \frac{(aq)_\infty}{(bq)_\infty} \sum_{n\geq 0} \frac{(b/a)_n}{(q)_n} \frac{(aq)^n}{1-tq^n},$$

and indeed this identity is easily verified, either directly, by expanding $(1-tq^n)^{-1}$, interchanging the order of summation, and taking into account (6.2), or by observing that the difference of the two members of (16.3) is an entire function of t that tends to zero as $|t| \to \infty$.

Similarly, by matching the singularities of $\phi(t) = (q)_\infty (1-t^{-1})^{-1} \cdot F(0, t^{-1}; t)$ at its poles $t = q^n$ $(-\infty < n < \infty)$, and comparing with

$$\psi(t) = \frac{1}{(1-t)(1-t^{-1})} + \sum_{n\geq 1} \frac{(-1)^n(1+q^n)q^{(3n^2+n)/2}}{(1-tq^n)(1-t^{-1}q^n)},$$

it is not difficult to show that $\phi(t) = \psi(t)$. Thus we have established

$$(16.4) \qquad (1-t)F(0, t^{-1}; t)$$
$$= \frac{1}{(q)_\infty} \left\{ 1 + (1-t)(1-t^{-1}) \sum_{n\geq 1} \frac{(-1)^n(1+q^n)q^{(3n^2+n)/2}}{(1-tq^n)(1-t^{-1}q^n)} \right\}.$$

17. Jacobi's triple product. Many proofs of this famous formula are known; we give one here that seems to fit in with the spirit of our approach. Consider

$$(17.1) \qquad g(a:q) = (-a^{-1}q)_\infty(-a)_\infty.$$

By (12.44) we may write

$$g(a:q) = \sum_{n\geq 0} \frac{q^{(n^2+n)/2}}{(q)_n} a^{-n} \sum_{m\geq 0} \frac{q^{(m^2-m)/2}}{(q)_m} a^m$$

$$= \sum_{N=-\infty}^{+\infty} a^N \sum_{\substack{m,n\geq 0 \\ m-n=N}} \frac{q^{(m^2+n^2+n-m)/2}}{(q)_m(q)_n}$$

$$= \sum_{N=-\infty}^{+\infty} a^N q^{(N^2-N)/2} \sum_{\substack{m,n\geq 0 \\ m-n=N}} \frac{q^{mn}}{(q)_m(q)_n}.$$

It is clear that the inner sum $S_N = S_{-N}$, so we need consider only nonnegative values of N. For such values we have

$$S_N = \sum_{n\geq 0} \frac{q^{n^2}q^{Nn}}{(q)_n(q)_{n+N}} = \frac{1}{(q)_N} \sum_{n\geq 0} \frac{q^{n^2}q^{Nn}}{(q)_n(q^{N+1})_n}$$

$$= \frac{1}{(q)_N} \cdot \frac{1}{(q^{N+1})_\infty} = \frac{1}{(q)_\infty},$$

the next to the last step being a consequence of (12.31). Hence

$$(17.2) \qquad (q)_\infty(-a^{-1}q)_\infty(-a)_\infty = \sum_{N=-\infty}^{+\infty} a^N q^{(N^2-N)/2}.$$

Replacing q by q^2 and then a by aq, we obtain the usual form

$$(17.3) \qquad \prod_{n \geq 1} (1 - q^{2n})(1 + aq^{2n-1})(1 + a^{-1}q^{2n-1}) = \sum_{N=-\infty}^{+\infty} a^N q^{N^2}.$$

An alternative approach is to derive a functional equation for the left side of (17.3) under $a \to aq^2$, assume a Laurent expansion with coefficients D_N, show that $D_N = C(q)q^{N^2}$, and evaluate $C(q)$ by taking $a = 1$ and using (7.325).

18. A bilateral series. We shall now extend the definition of $(aq)_n$ to negative values of n. With this extension we can then define a bilateral basic series. Our definition is by induction:

$$(18.1) \qquad (aq)_{n+1} = (1 - aq^{n+1})(aq)_n \qquad (-\infty < n < \infty).$$

It is easy to see that for $a \neq 0$, q^m ($m \geq 0$), and for all n,

$$(18.11) \qquad (aq)_{-n} = \frac{(-aq)^{-n}q^{(n^2+n)/2}}{(a^{-1})_n}.$$

Hence we may consider the function

$$(18.2) \qquad \begin{aligned} H(a,b;t) &\equiv \sum_{n=-\infty}^{+\infty} \frac{(aq)_n}{(bq)_n} t^n \\ &= F(a,b;t) + \sum_{n \geq 1} \frac{(b^{-1})_n}{(a^{-1})_n} (b/at)^n \\ &= F(a,b;t) + F((bq)^{-1}, (aq)^{-1}; b/at) - 1. \end{aligned}$$

The possible singularities of H as a function of t are poles at $t = q^{-m}$, $(b/a)q^m$ ($m \geq 0$), with (in general) essential singularities at $t = 0, \infty$.

From (18.1) and (18.2) it is clear that H satisfies the equation

$$(18.21) \qquad (1 - t)H(a,b;t) = b(1 - atq/b)H(a,b;tq).$$

Now, defining the auxiliary function

$$(18.22) \qquad G(a,b;t) \equiv G(t) \equiv H(a,b;t) \prod_{m \geq 0} (1 - tq^m)\left(1 - \frac{b}{at}q^m\right),$$

we have

$$(18.23) \qquad G(t) = -atqG(tq).$$

The only singularity of $G(t)$ in the finite plane is at $t = 0$. Hence, for $t \neq 0$, we may write

$$(18.24) \qquad G(t) = \sum_{n=-\infty}^{+\infty} B_n t^n,$$

and from (18.23),

$$(18.25) \qquad B_n = (-a)^n q^{(n^2+n)/2)} B_0 \qquad (-\infty < n < \infty).$$

Hence

$$G(t) = B_0 \sum_{n=-\infty}^{+\infty} (-at)^n q^{(n^2+n)/2}$$

(18.26)

$$= B_0 \prod_{m \geq 1} (1 - q^m)(1 - a^{-1}t^{-1}q^{m-1})(1 - atq^m),$$

the second form following from (17.2). To determine B_0, we observe that $F(b^{-1}q^{-1}, a^{-1}q^{-1}; b/at)$ is regular at $t = 1$, unless $ab^{-1} = q^m$, $m \geq 0$. Therefore (by (18.26) and (18.22))

$$G(1) = B_0 \prod_{m \geq 1} (1 - q^m)(1 - a^{-1}q^{m-1})(1 - aq^m)$$

$$= \prod_{m \geq 1} (1 - q^m) \left(1 - \frac{b}{a}q^{m-1}\right) \cdot \lim_{t \to 1} (1 - t)H(a, b; t)$$

$$= \prod_{m \geq 1} (1 - q^m) \left(1 - \frac{b}{a}q^{m-1}\right) \cdot \lim_{t \to 1} (1 - t)F(a, b; t)$$

$$= \prod_{m \geq 1} \frac{(1 - q^m)(1 - ba^{-1}q^{m-1})(1 - aq^m)}{(1 - bq^m)}.$$

Thus we have

(18.27) $$G(t) = \prod_{n \geq 1} \frac{(1 - q^n)(1 - a^{-1}t^{-1}q^{n-1})(1 - atq^n)(1 - ba^{-1}q^{n-1})}{(1 - a^{-1}q^{n-1})(1 - bq^n)},$$

or, reverting to $H(t)$ and our standard notation,

(18.3) $$H(a, b; t) = \frac{(q)_\infty (b/a)_\infty (a^{-1}t^{-1})_\infty (atq)_\infty}{(bq)_\infty (a^{-1})_\infty (t)_\infty (b/at)_\infty}.$$

This formula is given by Ramanujan. It contains as special cases many interesting identities and has numerous applications.

In (18.3), let $a = bq^{-1}$. The left side, by definition, becomes

$$H(bq^{-1}, b; t) = F(bq^{-1}, b; t) + F(b^{-1}q^{-1}, b^{-1}; qt^{-1}) - 1$$

(18.4)

$$= 1 + (1 - b) \sum_{n \geq 1} \frac{t^n}{1 - bq^n} - b^{-1}(1 - b) \sum_{n \geq 1} \frac{(qt^{-1})^n}{1 - b^{-1}q^n},$$

the two series converging for $|q| < |t| < 1$. If, in addition, $|q| < |b| < |q|^{-1}$, we may expand the denominators to get the double series

(18.5) $$H(bq^{-1}, b; t) = \frac{1 - bt}{1 - t} \left\{ 1 + \frac{(1 - b)(1 - t)}{(1 - bt)} \sum_{n,k \geq 1} a^{kn}(b^k t^n - b^{-k}t^{-n}) \right\}.$$

After simplifying the right side of (18.3) for $a = bq^{-1}$ and equating with (18.5), we get the identity

$$1 + \frac{(1 - b)(1 - t)}{(1 - bt)} \sum_{n,k \geq 1} q^{kn}(b^k t^n - b^{-k}t^{-n})$$

(18.6)

$$= \frac{(q)_\infty^2 (btq)_\infty (b^{-1}t^{-1}q)_\infty}{(bq)_\infty (b^{-1}q)_\infty (tq)_\infty (t^{-1}q)_\infty}.$$

The range of validity of (18.6) is easily seen to contain the region $|q| < |b| < |q|^{-1}$, $|q| < |t| < |q|^{-1}$. In particular, if we let $b = e^{iu}$, $t = e^{iv}$, where u and v are real, we obtain essentially a formula from the theory of elliptic functions:

(18.61)
$$\prod_{n \geq 1} \frac{(1 - q^n)^2 (1 - 2\cos(u+v)q^n + q^{2n})}{(1 - 2q^n \cos u + q^{2n})(1 - 2q^n \cos v + q^{2n})}$$

$$= 1 + 4 \frac{\sin(u/2)\sin(v/2)}{\sin((u+v)/2)} \sum_{N \geq 1} q^N \sum_{\substack{nk=N \\ n,k \geq 1}} \sin(ku + nv).$$

Specializing still further, we let $v = -u + \varepsilon$ and make $\varepsilon \to 0$. We get

(18.62)
$$\prod_{n \geq 1} \frac{(1 - q^n)^4}{(1 - 2q^n \cos u + q^{2n})^2}$$

$$= 1 - 8\sin^2 \frac{u}{2} \sum_{N \geq 1} q^N \sum_{\substack{nk=N \\ n,k \geq 1}} n \cos(k-n)u.$$

A comparison of (18.62) and (9.3) yields the identity

(18.7)
$$\left\{ 1 - 4\sin u \sum_{N \geq 1} q^N \sum_{\omega | N} \sin\left(\frac{2N}{\omega} - \omega\right) u \right\}^2$$

$$= 1 - 8\sin^2 u \sum_{N \geq 1} q^N \sum_{\substack{nk=N \\ n,k \geq 1}} n \cos 2(k-n)u.$$

For $b \neq 0, \infty$, q^n $(-\infty < n < \infty)$, define

(18.8) $\qquad J_b(t) = \dfrac{t}{(1-b)(1-b^{-1}q)} H(bq^{-1}, b; t) H(b^{-1}, qb^{-1}; t).$

The product expansion

$$\frac{t(1-bt)(1-bt^{-1})}{(1-b)^2(1-t)^2} \cdot \frac{(q)_\infty^4 (btq)_\infty (b^{-1}t^{-1}q)_\infty (bt^{-1}q)_\infty (b^{-1}tq)_\infty}{((bq)_\infty (b^{-1}q)_\infty (tq)_\infty (t^{-1}q)_\infty)^2}$$

shows that $J_b(t)$ satisfies the following conditions:

(18.81)
    (a) $J_b(tq) = J_b(t)$.
    (b) $J_b(t^{-1}) = J_b(t)$.
    (c) $J_b(t)$ is regular except at $t = 0, \infty, q^n$ $(-\infty < n < \infty)$.
    (d) $(1-t)^2 J_b(t) \to 1$ as $t \to 1$.

It is easy to see that conditions (18.81) determine $J_b(t)$ up to an additive constant. But the function

(18.82)
$$Z(t) = \sum_{n=-\infty}^{+\infty} \frac{tq^n}{(1 - tq^n)^2}$$

$$= \sum_{n \geq 0} \frac{tq^n}{(1 - tq^n)^2} + \sum_{n \geq 1} \frac{t^{-1}q^n}{(1 - t^{-1}q^n)^2}$$

also satisfies (18.81), so

$$J_b(t) = Z(t) - A(b).$$

But, again from the product, $J_b(t) = -J_t(b)$, so that $J_t(t) = 0$ and

(18.83) $$J_b(t) = Z(t) - Z(b).$$

This is essentially the formula which expresses the difference of two Weierstrass $\wp$-functions as an infinite product, and it has been applied by Bailey [6] to treat certain of Ramanujan's congruences. For $|q| < |t| < |q|^{-1}$, we may write

(18.84) $$Z(t) = \frac{t}{(1-t)^2} + \sum_{n,k \geq 1} q^{kn} k(t^k + t^{-k}),$$

and if we take $t = \omega^a$, $b = \omega^s$, $a \not\equiv \pm s \pmod p$, $\omega = \exp(2\pi i/p)$, (18.83) appears in the form

(18.85)
$$\prod_{n \geq 1} \frac{(1-q^n)^4(1-\omega^{a+s}q^n)(1-\omega^{-a-s}q^n)(1-\omega^{a-s}q^n)(1-\omega^{s-a}q^n)}{[(1-\omega^a q^n)(1-\omega^{-a}q^n)(1-\omega^s q^n)(1-\omega^{-s}q^n)]^2}$$
$$= 1 + A \sum_{n \geq 1} \frac{nq^n}{1-q^n} \left(\cos \frac{2\pi sn}{p} - \cos \frac{2\pi an}{p} \right),$$

where

$$A = 2\omega^{-s} \frac{(1-\omega^s)^2(1-\omega^a)^2}{(1-\omega^{a+s})(1-\omega^{a-s})}.$$

If we divide (18.83) by $(t-b)$ and let $b \to t$, we obtain, for $|q| < |t| < |q|^{-1}$,

(18.86) $$\frac{(q)_\infty^6 (t^{-2}q)_\infty (t^2)_\infty}{\{(t^{-1}q)_\infty (t)_\infty\}^4} = \frac{1+t}{(1-t)^3} + t^{-1} \sum_{n,k \geq 1} q^{kn} k^2 (t^k - t^{-k}),$$

a formula utilized by Carlitz [8]. Putting $t = \omega^a$ as above, we get

(18.87) $$\prod_{n \geq 1} \frac{(1-q^n)^6(1-q^n\omega^{2a})(1-q^n\omega^{-2a})}{[(1-q^n\omega^a)(1-q^n\omega^{-a})]^4} = 1 + B \sum_{n \geq 1} \frac{n^2 q^n}{1-q^n} \sin \frac{2\pi an}{p},$$

where $B = 2i\omega^{-a}(1-\omega^a)^4(1-\omega^{2a})^{-1}$.

19.   Two product-series identities. In (18.6), divided through by $(1-b)(1-t)(1-bt)^{-1}$, replace q by q^p, b by q^r, t by αq^s, where p is any integer greater than 2, r and s are positive integers satisfying $r + s < p$, $(r, p) = 1$, and α is a primitive kth root of unity, $\alpha = \exp(2\pi i a/k)$, $(a, k) = 1$, $k \geq 1$. The right side of (18.6) becomes

(19.1) $$\prod_{n \geq 1} \frac{(1-q^{pn})^2(1-\alpha^{-1}q^{pn-(r+s)})(1-\alpha q^{pn-(p-r-s)})}{(1-q^{pn-r})(1-q^{pn-p+r})(1-\alpha^{-1}q^{pn-s})(1-\alpha q^{pn-p+s})}.$$

The left side becomes

(19.2) $$\sum_{m,n \geq 0} q^{pmn+rm+sn} \alpha^n - \sum_{m,n > 0} q^{pmn-rm-sn} \alpha^{-n} = \sum_{N \geq 0} A(N) q^N,$$

where

$$A(N) = \sum_{\substack{pN+rs=(pm+s)(pn+r)\\m,n\geq 0}} \alpha^n - \sum_{\substack{pN+rs=(pm-s)(pn-r)\\m,n\geq 0}} \alpha^{-n}.$$

The sums run over all factorizations of $M = pN + rs$ into $d\delta$, where $d \equiv r$, $\delta \equiv s$ (mod p) in the first sum, $d \equiv -r$, $\delta \equiv -s$ (mod p) in the second. Since $(r,p) = 1$, $d \equiv r$ (mod p) implies $\delta \equiv s$ (mod p), and similarly for the second pair. Hence

(19.3) $$A(N) = \sum_{d|M} \{\chi_r(d;p)\alpha^{(d-r)/p} - \chi_{-r}(d;p)\alpha^{-(d+r)/p}\},$$

where $\chi_r(d;p) = 1$ if $d \equiv r$ (mod p), $= 0$ otherwise. Therefore

$$A(N) = \sum_{m=0}^{k-1} \alpha^m \left\{ \sum_{\substack{d|M\\(d-r)/p\equiv m(k)}} \chi_r(d;p) - \sum_{\substack{d|M\\-(d+r)/p\equiv m(k)}} \chi_{-r}(d;p) \right\}$$

$$= \sum_{m=0}^{k-1} \alpha^m \left\{ \sum_{\substack{d|M\\d\equiv mp+r(kp)}} - \sum_{\substack{d|M\\d\equiv -mp-r(k)}} 1 \right\}$$

$$= \sum_{m=0}^{k-1} \alpha^m E_{mp+r}(M; kp).$$

Thus we have established the identity

(19.4)
$$\prod_{n\geq 1} \frac{(1-q^{pn})^2(1-\alpha^{-1}q^{pn-(r+s)})(1-\alpha q^{pn-(p-r-s)})}{(1-q^{pn-r})(1-q^{pn-p+r})(1-\alpha^{-1}q^{pn-s})(1-\alpha q^{pn-p+s})}$$

$$= \sum_{N\geq 0} q^N \sum_{m=0}^{k-1} \alpha^m E_{mp+r}(pN + rs; kp).$$

By a slight modification of the derivation of (19.4) with $t = \alpha$ ($s = 0$), we can obtain the following identity, where $p \geq 2$:

(19.5)
$$\prod_{n\geq 1} \frac{(1-q^{pn})^2(1-\alpha^{-1}q^{pn-r})(1-\alpha q^{pn-p+r})}{(1-q^{pn-r})(1-q^{pn-p+r})(1-\alpha^{-1}q^{pn})(1-\alpha q^{pn})}$$

$$= 1 + (1-\alpha) \sum_{N\geq 1} q^N \sum_{m=0}^{k-1} \alpha^m E_{mp+r}(N; kp).$$

For $p = r + s$, (19.4) becomes

(19.6)
$$\prod_{n\geq 1} \frac{(1-q^{pn})^2(1-\alpha q^{pn})(1-\alpha^{-1}q^{pn})}{(1-q^{pn-r})(1-q^{pn-p+r})(1-\alpha q^{pn-r})(1-\alpha^{-1}q^{pn-p+r})}$$

$$= \sum_{N\geq 0} q^N(1-\alpha^{-1})^{-1} \sum_{m=0}^{k-1} \alpha^m E_{mp+r}(pN + r(p-r); kp).$$

For $p = 2$, this reduces to

$$
\prod_{n \geq 1} \frac{(1 - q^{2n})^2 (1 - 2\cos(2\pi a/k)q^{2n} + q^{4n})}{(1 - q^{2n-1})^2 (1 - 2\cos(2\pi a/k)q^{2n-1} + q^{4n-2})}
$$

(19.61)

$$
= \sum_{N \geq 0} q^N \sum_{0 \leq m < (k-1)/2} \frac{\sin(2m+1)(a\pi/k)}{\sin(a\pi/k)} E_{2m+1}(2N+1; 2k).
$$

20. A general transformation. Given a sequence A_n, $n \geq 0$, such that

(20.1)
$$
g(t) = \sum_{n \geq 0} A_n t^n
$$

converges, we define the linear transformation

(20.11)
$$
T_b(g) = h(t) = \sum_{n \geq 0} \frac{A_n}{(bq)_n} t^n.
$$

We shall show that $h(t)$ may be represented as a series in $g(t)$, $g(qt), \ldots$, with coefficients that depend on b in a simple way. In fact, for $g(t) = t^n$ $(n \geq 0)$,

$$
T_b(g) = \frac{t^n}{(bq)_n} = (bq)_\infty^{-1}(bq^{n+1})_\infty t^n
$$

$$
= (bq)_\infty^{-1} \sum_{r \geq 0} (-bq^n)^r \frac{q^{(r^2+r)/2}}{(q)_r} t^n,
$$

by using (12.44). This may be written in the form

(20.12)
$$
T_b(g) = (bq)_\infty^{-1} \sum_{r \geq 0} (-b)^r \frac{q^{(r^2+r)/2}}{(q)_r} g(q^r t).
$$

Equation (20.12), proved for the special case, is valid in general, by the linearity of T_b, provided that the series (20.11) converges. In all our applications this will be easily verified.

Choose $A_n = (aq)_n/(q)_n$, so that

$$
g(t) = F(a, 1; t) = \frac{(atq)_\infty}{(t)_\infty};
$$

then

$$
g(q^r t) = \frac{(t)_r}{(atq)_r} g(t),
$$

and (20.12) yields

(20.2)
$$
\sum_{n \geq 0} \frac{(aq)_n t^n}{(q)_n (bq)_n} = \frac{(atq)_\infty}{(bq)_\infty (t)_\infty} \sum_{r \geq 0} \frac{(t)_r}{(q)_r (atq)_r} (-b)^r q^{(r^2+r)/2}.
$$

This contains some of our earlier results as well as some new ones. For example, with $a = 0$ and t replaced by qt, we get

(20.21)
$$
\sum_{n \geq 0} \frac{(qt)^n}{(q)_n (bq)_n} = \frac{1}{(bq)_\infty (tq)_\infty} \sum_{r \geq 0} (-b)^r \frac{(tq)_r}{(q)_r} q^{(r^2+r)/2}.
$$

Setting $b = -1$, the left side becomes

$$F(0, 1; qt : q^2) = (tq; q^2)_\infty^{-1},$$

so

(20.22) $$\sum_{r \geq 0} \frac{(tq)_r}{(q)_r} q^{(r^2+r)/2} = (-q)_\infty (tq^2; q^2)_\infty = \prod_{n \geq 1} \frac{(1 - tq^{2n})}{(1 - q^{2n-1})}.$$

For $t = 1$, we recover (7.321); $t = q^{2k}$ and $t = q^{2k+1}$ yield (8.5a) and (8.5b). For $t = q^{-1/2}$, with q replaced by q^2, we find

(20.23)
$$1 + \frac{(1-q)}{(1-q^2)}q^2 + \frac{(1-q)(1-q^3)}{(1-q^2)(1-q^4)}q^6 + \frac{(1-q)(1-q^3)(1-q^5)}{(1-q^2)(1-q^4)(1-q^6)}q^{12} + \cdots$$
$$= \prod_{n \geq 1} \frac{(1 - q^{4n-1})}{(1 - q^{4n-2})}.$$

If we put $t = q$ in (20.2), we recover (7.3) essentially.

Suppose we write $a = t^{-1}c$ in (20.2) and let $t \to 0$. We obtain easily

(20.24)
$$(bq)_\infty \sum_{n \geq 0} (-c)^n \frac{q^{(n^2+n)/2}}{(q)_n (bq)_n}$$
$$= (cq)_\infty \sum_{n \geq 0} (-b)^n \frac{q^{(n^2+n)/2}}{(q)_n (cq)_n}.$$

Thus the function on the left is symmetric in b and c.

It is easy to invert the transformation T_a. Taking $h(t) = t^n$, $n \geq 0$, we have

$$T_a^{-1}(h) = g(t) = (aq)_n t^n = (aq)_\infty (aq^{n+1})_\infty^{-1} t^n$$
$$= (aq)_\infty \sum_{r \geq 0} \frac{(aq^{n+1})^r}{(q)_r} t^n,$$

(20.3) $$T_a^{-1}(h(t)) = (aq)_\infty \sum_{r \geq 0} \frac{(aq)^r}{(q)_r} h(q^r t).$$

Again, (20.3) is valid for general $h(t)$ by linearity.

By applying T_b followed by T_a^{-1}, we get

$$T_a^{-1} T_b(g(t)) = \sum_{n \geq 0} \frac{(aq)_n}{(bq)_n} A_n t^n = \frac{(aq)_\infty}{(bq)_\infty} \sum_{r,s \geq 0} \frac{(aq)^r (-b)^s}{(q)_r (q)_s} q^{(s^2+s)/2} g(q^{r+s} t)$$
$$= \frac{(aq)_\infty}{(bq)_\infty} \sum_{k \geq 0} \frac{(aq)^k}{(q)_k} g(q^k t) \sum_{\substack{r+s=k \\ r,s \geq 0}} \begin{bmatrix} k \\ s \end{bmatrix} \left(-\frac{b}{aq} \right)^s q^{(s^2+s)/2}.$$

Using (6.23) for the inner sum, we have

(20.4) $$T_a^{-1} T_b(g(t)) = \frac{(aq)_\infty}{(bq)_\infty} \sum_{k \geq 0} \frac{(b/a)_k}{(q)_k} (aq)^k g(q^k t).$$

If we apply (20.4) with $A_n = (cq)_n/(q)_n$, we obtain, as before,

(20.41) $$\sum_{n\geq 0} \frac{(aq)_n(cq)_n}{(q)_n(bq)_n} t^n = \frac{(aq)_\infty(ctq)_\infty}{(bq)_\infty(t)_\infty} \sum_{n\geq 0} \frac{(b/a)_n(t)_n}{(q)_n(ctq)_n}(aq)^n.$$

Hence, if we define

(20.42) $$F\left(\begin{matrix} a, c \\ b \end{matrix} ;t\right) = (tq)_\infty(bq^2)_\infty \sum_{n\geq 0} \frac{(aq)_n(cq)_n}{(q)_n(bq^2)_n}(qt)^n,$$

equation (20.41), with b, t replaced by bq, tq becomes

(20.43) $$F\left(\begin{matrix} a, c \\ b \end{matrix} ;t\right) = F\left(\begin{matrix} ba^{-1}, t \\ ct \end{matrix} ;a\right).$$

That is, the function (20.42) is invariant under the substitution

(20.44) $$a' = ba^{-1}, \quad b' = ct, \quad c' = t, \quad t' = a.$$

Thus, if we write $a = q^A$, $b = q^B$, $c = q^C$, $t = q^D$, (20.44) becomes a linear homogeneous substitution

(20.45)
$$S: \begin{matrix} A' = -A + B, \\ B' = \quad\quad C + D, \\ C' = \quad\quad D \\ D' = A. \end{matrix}$$

Because of the obvious symmetry in A and C, F is also invariant under

(20.46)
$$V: \begin{matrix} A' = \quad\quad C, \\ B' = \quad B, \\ C' = A, \\ D' = \quad\quad D. \end{matrix}$$

The substitutions S and V generate a group of order 12, with the relations

(20.47) $$S^6 = V^2 = (SV)^2 = I.$$

The transforms of the vector (A, B, C, D) under the group are: (20.48)

$I : (A, B, C, D),$ $V : (C, B, A, D),$

$S : (B - A, C + D, D, A),$ $VS : (D, C + D, B - A, A),$

$S^2 : (C + D - B + A, A + D, A, B - A),$ $VS^2 : (A, A + D, C + D - B + A, B - A),$

$S^3 : (B - C, B, B - A, C + D - B + A),$ $VS^3 : (B - A, B, B - C, C + D - B + A),$

$S^4 : (C, C + D, C + D - B + A, B - C),$ $VS^4 : (C + D - B + A, C + D, C, B - C),$

$S^5 : (D, A + D, B - C, C),$ $VS^5 : (B - C, A + D, D, C).$

Now if $A = -1$, F reduces to a simple product, and the same is true, therefore, if any one of the first elements in (20.48) is -1. For example, from S^2, with $C + D - B + A = -1$, we get (after $a \to aq^{-1}$, $b \to bq^{-1}$, $c \to cq^{-1}$)

(20.49) $$\sum_{n\geq 0} \frac{(a)_n(c)_n}{(q)_n(bq)_n} \left(\frac{bq}{ac}\right)^n = \frac{(bq/a)_\infty(bq/c)_\infty}{(bq)_\infty(bq/ac)_\infty}.$$

This summation formula is well known.

We can also derive the transformation (6.3) from

$$F\left(\begin{matrix} a,1 \\ bq^{-1} \end{matrix} ; tq^{-1}\right) = F\left(\begin{matrix} 1, at/b \\ tq^{-1} \end{matrix} ; bq^{-1}\right),$$

obtained by S^4 with $C = 0$. There are many interesting results which may be derived from (20.48), by applying the transformations to special cases. For example, if we start with

(20.5)
$$F\left(\begin{matrix} a,a \\ b \end{matrix} ; t\right) = F\left(\begin{matrix} b/a, t \\ at \end{matrix} ; a\right) = F\left(\begin{matrix} b/a, b/a \\ b \end{matrix} ; a^2 t/b\right)$$
$$= F\left(\begin{matrix} a, a^2 t/b \\ at \end{matrix} ; b/a\right),$$

by S, S^3, and $V S^2$ with $C = A$, and let $a \to 0$ in the first, second, and third, then replace b, t by bq^{-1}, tq^{-1}, we obtain

(20.51)
$$(t)_\infty (bq)_\infty \sum_{n \geq 0} \frac{t^n}{(q)_n (bq)_n}$$
$$= \sum_{n \geq 0} \frac{(t)_n}{(q)_n} (-b)^n q^{(n^2+n)/2} = (bq)_\infty \sum_{n \geq 0} \frac{(bt)^n q^{n^2}}{(q)_n (bq)_n}.$$

Similarly, if we let $b \to 0$ in the first, second, and fourth forms of (20.5), and then replace t by tq^{-1}, we get

(20.52)
$$(t)_\infty \sum_{n \geq 0} \frac{(aq)_n^2}{(q)_n} t^n$$
$$= (aq)_\infty (atq)_\infty \sum_{n \geq 0} \frac{(t)_n}{(q)_n (atq)_n} (aq)^n$$
$$= (atq)_\infty \sum_{n \geq 0} \frac{(aq)_n}{(q)_n (atq)_n} (-at)^n q^{(n^2+n)/2}.$$

A double application of the inverse transformation (20.3) yields

$$T_{a_1}^{-1} T_a^{-1}(g(t)) = (a, q)_\infty \sum_{k,r \geq 0} \frac{(a_1 q)^k (aq)^r}{(q)_k (q)_r} g(q^{k+r} t)$$
$$= (a_1 q)_\infty (aq)_\infty \sum_{n \geq 0} \frac{(aq)^n}{(q)_n} g(q^n t) \sum_{k=0}^{n} \begin{bmatrix} n \\ k \end{bmatrix} \left(\frac{a_1}{a}\right)^k.$$

Hence, if we define

(20.6)
$$f_n(u) = \sum_{k=0}^{n} \begin{bmatrix} n \\ k \end{bmatrix} u^k$$

and write $a_1 = au$,

(20.61)
$$T_{au}^{-1} T_a^{-1}(g(t)) = (aq)_\infty (auq)_\infty \sum_{n \geq 0} \frac{(aq)^n}{(q)_n} f_n(u) g(q^n t).$$

Certain special cases of the function $f_n(u)$ have been studied. We shall develop some of its properties here.

Let $t = 0$ in (20.61). Then

(20.62) $$\sum_{n\geq 0} \frac{f_n(u)}{(q)_n}(aq)^n = (aq)_\infty^{-1}(auq)_\infty^{-1} \equiv W(a, u).$$

Several functional equations for W are of interest. The first three, of course, are not independent.

(20.63a) $$W(aq, u) = (1 - aq)(1 - auq)W(a, u),$$

(20.64a) $$W(a, uq) = (1 - auq)W(a, u),$$

(20.65a) $$W(aq, u) = (1 - aq)W(a, uq),$$

(20.66a) $$W(a, u^{-1}) = W(au^{-1}, u).$$

By equating coefficients of $(aq)^n$, we get, respectively,

(20.63b) $$f_{n+1}(u) - (1 + u)f_n(u) + u(1 - q^n)f_{n-1}(u) = 0,$$

(20.64b) $$f_n(uq) - f_n(u) + u(1 - q^n)f_{n-1}(u) = 0,$$

(20.65b) $$q^n f_n(u) - f_n(uq) + (1 - q^n)f_{n-1}(uq) = 0,$$

(20.66b) $$f_n(u^{-1}) = u^{-n} f_n(u).$$

This last equation can be derived directly from (20.6).

Now $f_0(u) = 1$, $f_1(u) = 1 + u$, and if we set $u = -1$ in (20.63b), we get

$$f_{n+1}(-1) = (1 - q^n)f_{n-1}(-1).$$

Hence

(20.67) $$f_{2n}(-1) = (1 - q)(1 - q^3)\cdots(1 - q^{2n-1}) = (q; q^2)_n,$$
$$f_{2n+1}(-1) = 0.$$

This could have been established directly from (20.62), since

$$W(a, -1) = (a^2 q^2; q^2)_\infty^{-1} = \sum_{n\geq 0} \frac{(a^2 q^2)^n}{(q^2; q^2)_n}.$$

Also, the second part of (20.67) is a simple consequence of (20.66b).

Similarly, putting $u = q^{1/2}$ in (20.62), we get

$$\sum_{n\geq 0} \frac{f_n(q^{1/2})}{(q)_n}(aq)^n = \left\{\prod_{n\geq 1}(1 - aq^n)(1 - aq^{n+1/2})\right\}^{-1}$$

$$= \left\{\prod_{n\geq 2}(1 - aq^{n/2})\right\}^{-1} = (aq; q^{1/2})_\infty^{-1} = \sum_{n\geq 0} \frac{(aq)^n}{(q^{1/2}; q^{1/2})_n}.$$

Hence

$$
\begin{aligned}
(20.68) \quad f_n(q^{1/2}) &= \frac{(q)_n}{(q^{1/2}; q^{1/2})_n} \\
&= (1 + q^{1/2})(1 + q)(1 + q^{3/2}) \cdots (1 + q^{n/2}) = (-q^{1/2}; q^{1/2})_n.
\end{aligned}
$$

It is easy to obtain other values. For example, by (20.66b),

$$
(20.69) \qquad f_n(q^{-1/2}) = q^{-n/2} f_n(q^{1/2}).
$$

We can also derive the exact generating function for $f_n(u)$. For this purpose, set $A_n = q^n/(q)_n$ in (20.61), so that $g(t) = (tq)_\infty^{-1}$ and $g(q^n t) = (tq)_n \, g(t)$. Then

$$
\begin{aligned}
T_{au}^{-1} T_a^{-1}(g(t)) &= \sum_{n \geq 0} \frac{(aq)_n (auq)_n}{(q)_n} (qt)^n \\
&= (aq)_\infty (auq)_\infty (tq)_\infty^{-1} \sum_{n \geq 0} \frac{(tq)_n}{(q)_n} f_n(u)(aq)^n,
\end{aligned}
$$

and by definition (20.42), with $b = 0$, $c = au$,

$$
(aq)_\infty (auq)_\infty \sum_{n \geq 0} \frac{(tq)_n}{(q)_n} f_n(u)(aq)^n = F \begin{pmatrix} a, au \\ 0 \end{pmatrix}; t = F \begin{pmatrix} 0, t \\ aut \end{pmatrix}; a,
$$

the last by virtue of (20.43). Setting $t = 1$, simplifying, and replacing a by aq^{-1}, we get

$$
(20.7) \qquad \sum_{n \geq 0} f_n(u)a^n = (1 - au)^{-1} F(0, au; a) = (1 - a)^{-1} F(0, a; au).
$$

As an application, set $u = -1$ and use (20.67):

$$
\begin{aligned}
(1 + a)^{-1} F(0, -a; a) &= \sum_{n \geq 0} f_{2n}(-1)a^{2n} = \sum_{n \geq 0} (q; q^2)_n a^{2n} \\
&= F(q^{-1}, 0; a^2 : q^2).
\end{aligned}
$$

Thus

$$
(20.71) \qquad (1 + a)^{-1} F(0, -a; a : q) = F(q^{-1}, 0; a^2 : q^2).
$$

If we put $u = q^{1/2}$ in (20.7) and use (20.68), we get

$$
\begin{aligned}
(1 - a)^{-1} F(0, a; aq^{1/2}) &= \sum_{n \geq 0} f_n(q^{1/2})a^n \\
&= \sum_{n \geq 0} (-q^{1/2}; q^{1/2})_n a^n = F(-1, 0; a : q^{1/2}),
\end{aligned}
$$

and replacing q by q^2,

$$
(20.72) \qquad (1 - a)F(-1, 0; a : q) = F(0, a; aq : q^2).
$$

Both (20.71) and (20.72) appear to be new. The latter can be used to obtain transformations of some of the mock theta-functions of Ramanujan, and also to derive a theorem on partitions (23.91).

21. The basic multinomial. The function $f_n(u)$ defined in the preceding section is of some interest as a generalization of the binomial $(1+u)^n$, to which it reduces for $q = 1$. Similarly, we can define the homogeneous function of k variables

$$(21.1) \qquad Q_n(u_1, \ldots, u_k) = \sum_{\substack{r_1 + \cdots + r_k = n \\ r_i \geq 0}} \frac{(q)_n}{(q)_{r_1}(q)_{r_2} \cdots (q)_{r_k}} u_1^{r_1} \cdots u_k^{r_k}.$$

It is easy to see that

$$(21.11) \qquad \sum_{n \geq 0} \frac{Q_n(u_1, \ldots, u_k)}{(q)_n} a^n = \{(u_1 a)_\infty \cdots (u_k a)_\infty\}^{-1}.$$

Two special cases are of immediate interest. First, let $u_s = \exp(2\pi i s/k)$, $s = 0, 1, 2, \ldots, k - 1$. The product in (21.11) becomes

$$(a^k; q^k)_\infty^{-1} = \sum_{n \geq 0} \frac{a^{kn}}{(q^k; q^k)_n}.$$

Equating coefficients, we find

$$(21.2) \qquad \begin{aligned} Q_n(u_1, \ldots, u_k) &= 0 \quad \text{for } n \not\equiv 0 \ (\mathrm{mod}\ k), \\ Q_n &= \frac{(q)_n}{(q^k; q^k)_{n/k}} \quad \text{for } n \equiv 0 \ (\mathrm{mod}\ k). \end{aligned} \qquad (u_s = \exp(2\pi i s/k)).$$

Next, let $u_r = q^{(r-1)/p}$ $(r = 1, 2, \ldots, p)$. The product in (21.11) becomes

$$(a; q^{1/p})_\infty^{-1} = \sum_{n \geq 0} \frac{a^n}{(q^{1/p}; q^{1/p})_n},$$

so that, on equating coefficients, we get

$$(21.21) \qquad Q_n(1, q^{1/p}, q^{2/p}, \ldots, q^{(p-1)/p}) = \frac{(q)_n}{(q^{1/p}; q^{1/p})_n}.$$

A combination of these two special cases leads to a more general result. Let k, p be positive integers, and let

$$(21.22) \qquad u_{rs} = \exp(2\pi i s/k) q^{(r-1)/p} \qquad (r = 1, \ldots, p; \ s = 0, \ldots, k - 1).$$

With these kp arguments we find, as above

$$(21.221) \qquad \sum_{m \geq 0} \frac{Q_m}{(q)_m} a^m = (a^k; q^{k/p})_\infty^{-1} = \sum_{n \geq 0} \frac{a^{kn}}{(q^{k/p}; q^{k/p})_n}.$$

Hence

$$(21.23) \qquad \begin{aligned} Q_m &= 0 \quad \text{if } m \not\equiv 0 \ (\mathrm{mod}\ k), \\ Q_{nk} &= \frac{(q)_{nk}}{(q^{k/p}; q^{k/p})_n} \qquad (n \geq 0). \end{aligned}$$

Now the function Q_n is related to the N-fold application of T_u^{-1}. In fact

$$\begin{aligned} (21.3) \qquad T_{au_1}^{-1} T_{au_2}^{-1} \cdots T_{au_N}^{-1}(g(t)) &= h(t) \\ &= \sum_{n \geq 0} (au_1 q)_n \cdots (au_N q)_n A_n t^n \\ &= (au_1 q)_\infty \cdots (au_N q)_\infty \sum_{n \geq 0} \frac{Q_n}{(q)_n} (aq)^n g(tq^n). \end{aligned}$$

Take $\Lambda_n = (cq)_n/(q)_n$, so $g(t) = (ctq)_\infty/(t)_\infty$ and $g(tq^n) = g(t)(t)_n/(ctq)_n$. Then (21.3) becomes

(21.31)
$$\sum_{n\geq 0} \frac{(au_1 q)_n \cdots (au_N q)_n (cq)_n}{(q)_n} t^n$$
$$= \frac{(au_1 q)_\infty \cdots (au_N q)_\infty (ctq)_\infty}{(t)_\infty} \sum_{n\geq 0} \frac{Q_n}{(q)_n} \cdot \frac{(t)_n}{(ctq)_n} (aq)^n.$$

Now choose the u's as in (21.22). Then

$$\prod_{r,s}(au_{rs}q)_n = ((aq)^k; q^{k/p})_{np},$$

so (21.31) becomes, after using (21.23) and replacing a by $a^{1/k}q^{-1}$,

(21.32)
$$\sum_{n\geq 0} \frac{(a; q^{k/p})_{np}(cq)_n}{(q)_n} t^n = \frac{(a; q^{k/p})_\infty (ctq)_\infty}{(t)_\infty} \sum_{n\geq 0} \frac{(t)_{nk} a^n}{(q^{k/p}; q^{k/p})_n (ctq)_{nk}}.$$

Now replace $q^{k/p}$ by q_1. Then

(21.33)
$$\sum_{n\geq 0} \frac{(a; q_1)_{np}(cq)_n}{(q)_n} t^n$$
$$= \frac{(a; q_1)_\infty (ctq)_\infty}{(t)_\infty} \sum_{n\geq 0} \frac{(t)_{nk} a^n}{(q_1; q_1)_n (ctq)_{nk}}.$$

This identity in q and q_1, as it stands, has meaning only for $q_1 = q^{k/p}$. However, if we use

$$(b; q)_n = \frac{(b; q)_\infty}{(bq^n; q)_\infty}$$

it may be written as

(21.34)
$$\frac{(cq; q)_\infty}{(q; q)_\infty} \sum_{n\geq 0} \frac{(q^{n+1}; q)_\infty t^n}{(aq_1^{np}; q_1)_\infty (cq^{n+1}; q)_\infty}$$
$$= \frac{1}{(q_1; q_1)_\infty} \sum_{n\geq 0} \frac{(q_1^{n+1}; q_1)_\infty (ctq^{kn+1}; q)_\infty a^n}{(tq^{kn}; q)_\infty}.$$

Further, let $y = q_1^p = q^k$. Then we have (with $c \to cq^{-1}$)

(21.4)
$$\frac{(c; q)_\infty}{(q; q)_\infty} \sum_{n\geq 0} \frac{(q^{n+1}; q)_\infty t^n}{(ay^n; q_1)_\infty (cq^n; q)_\infty}$$
$$= \frac{1}{(q_1; q_1)_\infty} \sum_{n\geq 0} \frac{(q_1^{n+1}; q_1)_\infty (cty^n; q)_\infty a^n}{(ty^n; q)_\infty}.$$

Now we may consider the validity of (21.4) for independent values of q, q_1, y. Since the special values $q_1 = q^{k/p}$, $y = q^k$ are sufficiently numerous, an analytic continuation argument will work. Alternatively, we can expand the left side in powers of a and compare coefficients. Either way, (21.4) is established in full generality. It is an example of a *bi-basic* identity.

The methods of §20 can be pushed a little further. Suppose that, for $g(t) = \sum A_n t^n$, we define

$$(21.7) \qquad S(g) = h(t) = \sum_{n\geq 0} A_n \frac{(by^n)_\infty}{(ay^n)_\infty} t^n.$$

It is then fairly easy to show that

$$(21.8) \qquad h(t) = \sum_{n\geq 0} \frac{(b/a)_n}{(q)_n} a^n g(y^n t),$$

similarly to (20.4). Choosing $A_n = (cy; y)_n/(y; y)_n$, we get

$$(21.81) \qquad \begin{aligned} &\sum_{n\geq 0} \frac{(cy; y)_n}{(y; y)_n} \cdot \frac{(by^n)_\infty}{(ay^n)_\infty} t^n \\ &= \frac{(cty; y)_\infty}{(t; y)_\infty} \sum_{n\geq 0} \frac{(b/a)_n (t; y)_n}{(q; q)_n (cty; y)_n} a^n. \end{aligned}$$

After replacing c by cy^{-1} and making appropriate changes in notation, we find another bi-basic identity:

$$(21.9) \qquad \begin{aligned} &\frac{(c; y)_\infty}{(y; y)_\infty} \sum_{n\geq 0} \frac{(y^{n+1}; y)_\infty (by^n; q)_\infty}{(cy^n; y)_\infty (ay^n; q)_\infty} t^n \\ &= \frac{(b/a; q)_\infty}{(q; q)_\infty} \sum_{n\geq 0} \frac{(q^{n+1}; q)_\infty (cty^n; y)_\infty}{(bq^n/a; q)_\infty (ty^n; y)_\infty} a^n. \end{aligned}$$

Notes

§1. The most natural question that occurs in studying this book is: Why does

$$F(a, b; t : q) = \sum_{n=0}^{\infty} \frac{(aq)_n t^n}{(bq)_n}$$

have so much structure and yield such diverse and interesting results in such a natural way?

A vague philosophical question like this is impossible to answer fully. Indeed two other mathematicians, L. J. Rogers [26, pp. 334–335] and G. W. Starcher [28], both saw the importance of $F(a, b; t: q)$; however neither pursued its properties in anything like the exhaustive manner followed in this book. I feel something of the magic of $F(a, b; t:q)$ must lie in the following observation:

Let us consider a distribution of weights w_i at the points $x = q^i$ ($i = 0, 1, 2, \dots$), where $0 < q < 1$. The w_i are given by

$$w_i = \frac{a^i q^i (q^{i+1})_\infty}{(bq^i/a)_\infty}.$$

This is a q-analog of the classical β-distribution. The moment generating function for this distribution is:

$$\int \frac{d\omega}{1 - tx} = \sum_{i=0}^{\infty} \frac{a^i q^i (q^{i+1})_\infty}{(1 - tq^i)(bq^i/a)_\infty}$$

$$= \frac{(q)_\infty}{(b/a)_\infty} \sum_{i=0}^{\infty} \frac{a^i q^i (b/a)_i}{(1 - tq^i)(q)_i}$$

$$= \frac{(q)_\infty}{(b/a)_\infty} \sum_{m=0}^{\infty} t^m \sum_{i=0}^{\infty} \frac{a^i q^{(m+1)i} (b/a)_i}{(q)_i}$$

$$= \frac{(q)_\infty}{(b/a)_\infty} \sum_{m=0}^{\infty} \frac{t^m (bq^{m+1})_\infty}{(aq^{m+1})_\infty}$$

$$\text{(by (6.2) of Chapter 1)}$$

$$= \frac{(q)_\infty (bq)_\infty}{(b/a)_\infty (aq)_\infty} F(a, b; t : q).$$

The orthogonal polynomials $p_n(x)$ connected with this distribution are

$$p_n(x) = \sum_{j=0}^{n} \frac{(q^{-n})_j (bq^n)_j (xq)^j}{(q)_j (aq)_j}.$$

These are q-analogs of the Jacobi polynomials scaled to the interval $[0, 1]$. They were first given by W. Hahn [18], and many details about them appear in [15]. Since these polynomials can be specialized to all of the classical continuous orthogonal polynomials and since their moment generating function effectively encodes enough information to define them, we should not be surprised at the richness and importance of $F(a, b; t : q)$.

The basic properties of Gauss polynomials are given in Chapter 3 of [8].

§4. Equation (4.1) was given by L. J. Rogers [26, p. 334, (1)].

§6. Equation (6.1) was given by Starcher [28, (3.16) with $\lambda = 1$]. Equation (6.2) is simultaneously first the q-analog of the binomial series [18, (3.3)] and second the q-analog of the beta integral [18, (3.11)]. Thus this elementary result turns out to be immensely powerful in applications. Applications of (6.2) and its corollaries are given in [8, Chapter 2].

Equation (6.3) is an elegant involution with a simple combinatorial interpretation and proof [6, (5.1), pp. 574–575].

§7. The main result of this section is identity (7.2). It is equivalent to Theorem 1 of [10] with $B = 0$; actually equation (6.3) is required to then identify the resulting identity with (7.2). Thus a result of immense importance in much of Ramanujan's "Lost" Notebook is a natural component of this work. R. P. Agarwal [1] has generalized Theorem 1 of [10] and has shown that it follows from three-term relations for the more general $_3\phi_2$ basic hypergeometric function.

Equation (7.321) (due to Gauss) is easily derived from Jacobi's Triple Product Identity [8, p. 27, history, p. 31].

Equation (7.7) is due to L. J. Rogers [26, p. 335, (4)] and (7.8) is Euler's celebrated Pentagonal Number Theorem [8, p. 11].

§8. Identity (8.43) was derived in a different manner in [4, p. 137, Theorem F_2].

§9. and §10. Equation (9.3) is (as noted there) the expansion for the reciprocal of a theta function. It has played an important part in the elucidation of certain Hecke modular form type identities [11]. Historical references are given in [11].

§11. Identity (11.1) is equivalent to Theorem 1 of [10] with $A = 0$. Thus the comments on equation (7.2) are relevant here as well.

§12. Equation (12.2) is a special case of an identity due to F. H. Jackson [21, p. 145, (4)]. Jackson's result is reproved in [7, §3] and applied to several summation theorems. The statement of Jackson's identity in [7, p. 527, Lemma] reduces to (12.2) when $\alpha = q$. There is a combinatorial proof of (12.2) in [6, (5.4), pp. 575–576]. Equation (12.31) is probably due to Cauchy [8, pp. 20–21]. Equation (12.44) is due to Euler (cf. [8, p. 19 (2.26)]).

§13. This section seems more than any in Chapter 1 to be begging for further study. Identities (13.1), (13.2), and (13.5) do not seem to fit into the known q-hypergeometric hierarchy of results.

§14. Equation (14.1) is due to L. J. Rogers [26, p. 334,(1)]. A combinatorial proof of (14.1) was given in [6, §4], where it was christened the "Rogers-Fine Identity." It should be mentioned that it was also given by G. W. Starcher [28, p. 803, (1.12)] and that it plays a substantial role in proving many results in Ramanujan's "Lost" Notebook [10].

§15. Equation (15.5) is another result unlike most in the literature. It is somewhat reminiscent of [10, p. 156, (42)] and [13, Lemma 1]. As with §13 we note that such results are not currently embedded in the q-hypergeometric hierarchy.

§16. Identity (16.3) is the special case of Heine's transformation (see equation (20.41) with $c = 1$) [19, p. 106]. Equation (16.4) follows from Watson's q-analog of Whipple's theorem [29; 27, p. 100, (3,4.1.5)] and is equivalent (modulo (12.3)) to the formula given by Watson to treat the third-order mock theta functions [30, p. 64; 9, 3.3)].

§17. The proof of Jacobi's triple product given here is closely related to that given originally by Jacobi [22, §64]. There have been numerous proofs of this result over the years (see [2; 8, pp. 30–31]). Also the many combinatorial proofs are examined in [12], and the related technique is applied to a wide class of combinatorial objects.

§18. The main result of this section is (18.3), which is often referred to as "Ramanujan's $_1\psi_1$-summation." There are several proofs in the literature [5, 16, 18, 20]; of these the one by M. Ismail [20] is by far the most elegant and succinct. Ismail's proof is given in Appendix C of [14], and numerous applications of it appear throughout [14] especially in Chapters 4 and 5.

§20. In this section, Heine's transformation [**19**, p. 106] of the q-hypergeometric function is derived from the study of the linear operator T_b. Heine's transformation is (20.41). As an interesting corollary of this result, it is shown that Heine's transformation and the natural symmetry of upper parameters generate a group of order 12. This was apparently understood by L. J. Rogers [**23**, p. 171], who isolated the three central transformations of this group. Equation (20.49) was also deduced by Heine (cf. [**8**, p. 20]).

The section closes with a careful study of the polynomials $f_n(u)$ defined by (20.6). These polynomials have come to be called the Rogers-Szegö polynomials, and the reader is referred to [**8**, pp. 49–50] for a discussion of their properties and the relevant literature.

§21. The polynomials $f_n(u)$ of §20 are generalized in (21.1) to $Q_n(u_1, \ldots, u_k)$, which are a q-analog of the multinomial expansion. The $Q_n(u_1, \ldots, u_k)$ were extensively studied by L. J. Rogers [**24**] in his first memoir "On the Expansion of Some Infinite Products" and provided the foundations on which Rogers built up to the Rogers-Ramanujan identities [**25**]. The remainder of the section leads up to the identity (21.9), which is the case $k = r = 1$, $s = l = 0$ of the Fundamental Lemma in [**3**, p. 65]. It is clear from the results in Ramanujan's "Lost" Notebook [**9**, **10**] that he fully understood this technique also.

References

1. R. P. Agarwal, *On the paper "A 'lost' notebook of Ramanujan—Partial theta functions"* of G. E. Andrews, Adv. in Math. **53** (1984), 291–300.

2. G. E. Andrews, *A simple proof of Jacobi's triple product identity*, Proc. Amer. Math. Soc. **16** (1965), 333–334.

3. _____, *On basic hypergeometric functions, mock theta functions and partitions*. (I), Quart. J. Math. Oxford Ser. (2) **17** (1966), 64–80.

4. _____, *On basic hypergeometric functions, mock theta functions and partitions*. (II), Quart J. Math. Oxford Ser. (2) **17** (1966), 132–143.

5. _____, *On Ramanujan's summation of $_1\psi_1 (a; b; z)$*, Proc. Amer. Math. Soc. **22** (1969), 552–553.

6. _____, *Two theorems of Gauss and allied identities proved arithmetically*, Pacific J. Math. **41** (1972), 563–568.

7. _____, *On the q-analog of Kummer's theorem and applications*, Duke Math. J. **40** (1973), 525–528.

8. _____, *The theory of partitions*, Encyclopedia of Mathematics and Its Applications, Vol. 2, G.-C. Rota editor, Addison-Wesley, Reading, Mass.,1976; reprinted by Cambridge Univ. Press, 1984.

9. _____, *An introduction to Ramanujan's 'lost' notebook*, Amer. Math. Monthly **86** (1979), 89–108.

10. _____, *Ramanujan's 'lost' notebook. I. Partial ϑ-functions*, Adv. in Math. **41** (1981), 137–172.

11. _____, *Hecke modular forms and the Kac-Peterson identities*, Trans. Amer. Math. Soc. **283** (1984), 451–458.

12. _____, *Generalized Frobenius partitions*, Mem. Amer. Math. Soc. vol. 49, No. 301 (1984).

13. _____, *Ramanujan's 'lost' notebook. IV. Stacks and alternating parity in partitions*, Adv. in Math. **53** (1984), 55–74.

14. ____, *q-Series: Their development and application in analysis, number theory, combinatorics, physics and computer algebra*, CBMS Regional Conf. Ser. in Math., no. 66, Amer. Math. Soc., Providence, R.I., 1986.

15. G. E. Andrews and R. Askey, *Enumeration of partitions: The role of Eulerian series and q-orthogonal polynomials*, Higher Combinatorics, M. Aigner, editor, Reidel, Dordrecht, Holland, 1977.

16. ____, *A simple proof of Ramanujan's $_1\psi_1$*, Aequationes Math. **18** (1978), 333–337.

17. R. Bellman, *A brief introduction to theta functions*, Holt, Rinehart and Winston, New York, 1961.

18. W. Hahn, *Über Orthogonalpolynome, die q-Differenzengleichungen genügen*, Math. Nach. **2** (1949), 4–34.

19. E. Heine, *Handbuch der Kugelfunktionen*, Vol. 1, Reimer, Berlin, 1878; reprinted by Physica-Verlag, Wurzburg, 1961.

20. M. Ismail, *A simple proof of Ramanujan's $_1\psi_1$ sum*, Proc. Amer. Math. Soc. **63** (1977), 185–186.

21. F. H. Jackson, *Transformations of q-series*, Mess. of Math. **39** (1910), 145–153.

22. C. G. J. Jacobi, *Fundamenta nova theoriae funktionum ellipticarum*, Regiomonti fratrum Bornträger,1829; reprinted in Gesammelte Werke, Vol. 1, Reimer, Berlin, 1881, pp. 49–239.

23. L. J. Rogers, *On a three-fold symmetry in the elements of Heine's series*, Proc. London Math. Soc. **24** (1893), 171–179.

24. ____, *On the expansion of some infinite products*, Proc. London Math. Soc. **24** (1893), 337–352.

25. ____, *Second memoir on the expansion of some infinite products*, Proc. London Math. Soc. **25** (1894), 318–343.

26. ____, *On two theorems of combinatory analysis and some allied identities*, Proc. London Math. Soc., Ser 2 **16** (1916), 315–336.

27. L. J. Slater, *Generalized hypergeometric functions*, Cambridge Univ. Press, 1966.

28. G. W. Starcher, *On identities arising from solutions of q-difference equations and some interpretations in number theory*, Amer. J. Math. **53** (1930), 801–816.

29. G. N. Watson, *A new proof of the Rogers-Ramanujan identities*, J. London Math. Soc. **4** (1929), 4–9.

30. ____, *The final problem: An account of the mock-theta functions*, J. London Math. Soc. **11** (1936), 55–80.

CHAPTER 2

Partitions

22.   Sums over partitions. We define a *partition* as a sequence $\pi = (k_1, k_2, \dots)$ of nonnegative integers, with the single restriction that all but a finite number of the k_j are zero. The *content* of π, $c(\pi)$, is defined by

$$(22.1) \qquad\qquad c(\pi) = \sum j k_j.$$

We say that π is a *partition of* $n = c(\pi)$. We may think of a partition of n as a "way" of writing n as a sum of positive integers j, each one occurring with frequency or multiplicity k_j, order being immaterial. An alternate notation is $\pi = (1^{k_1} 2^{k_2} \cdots)$, where only the positive frequencies k_j are mentioned. Thus the partition $(1, 0, 2, 0, 0, \dots)$ with content 7 would be written $\pi = (1^1 3^2)$.

We shall be dealing with sums of the form $\sum f(\pi)$, the summation being extended over all partitions of n. (We indicate this by writing $\sum_{\pi(n)} f(\pi)$.)

Two particular functions that we shall discuss are

$$(22.11) \qquad\qquad k(\pi) = \sum_j k_j,$$

the total number of parts in π, and

$$(22.12) \qquad\qquad Q(\pi) = \sum_{k_j > 0} 1,$$

the number of distinct parts in π. For example, the partition $\pi(1^1 3^2)$ of 7 has $k_1 = 1$, $k_2 = 0$, $k_3 = 2$, $k_4 = \cdots = 0$, and $k(\pi) = 3$, $Q(\pi) = 2$.

We begin with an important though obvious remark.

THEOREM 1.   *Let* $\psi_j(q) = \sum_{n \geq 0} C_j(k) q^k$ $(j = 1, 2, \dots)$. *Then*

$$\prod_{j \geq 1} \psi_j(q^j) = \sum_{n \geq 0} q^n \sum_{\pi(n)} C_1(k_1) C_2(k_2) \cdots.$$

There is no question of convergence here, since the power series and their product are purely formal.

We give some examples.

EXAMPLE 1. $C_j(k) = 1$ $(j \geq 1,\ k \geq 0)$. Then $\psi_j(q) = (1 - q)^{-1}$,

$$(22.13) \qquad \prod_{j \geq 1} (1 - q^j)^{-1} = P(q) = \sum_{n \geq 0} q^n \sum_{\pi(n)} 1 = \sum_{n \geq 0} p(n) q^n,$$

where $p(n)$ is the unrestricted partition function.

37

EXAMPLE 2. $C_j(k) = (-1)^k$ $(j \geq 1,\ k \geq 0)$. Then $\psi_j(q) = (1+q)^{-1}$,

$$(22.14) \qquad \prod_{j \geq 1}(1+q^j)^{-1} = \sum_{n \geq 0} q^n \sum_{\pi(n)} (-1)^{k_1+k_2+\cdots} = \sum_{n \geq 0} q^n \sum_{\pi(n)} (-1)^{k(\pi)}.$$

This is the generating function for $p_E(n) - p_O(n)$ where $p_E(n)$ is the number of partitions of n into an even number of parts, $p_O(n) = p(n) - p_E(n)$.

EXAMPLE 3. $C_j(k) = u^k$ $(j \geq 1,\ k > 0)$. Then

$$(22.15) \qquad \prod_{j \geq 1}(1-uq^j)^{-1} = \sum_{n \geq 0} q^n \sum_{\pi(n)} u^{k(\pi)}.$$

EXAMPLE 4. $C_j(k) = u_j^k$ $(j \geq 1,\ k \geq 0)$. Then

$$(22.16) \qquad \prod_{j \geq 1}(1-u_j q^j)^{-1} = \sum_{n \geq 0} q^n \sum_{\pi(n)} u_1^{k_1} u_2^{k_2} \cdots.$$

EXAMPLE 5. $\psi_j(q) = e^{tq}$ $(j \geq 1)$. Then $C_j(k) = t^k/k!$, so

$$\sum_{n \geq 0} q^n \sum_{\pi(n)} \frac{t^{k(\pi)}}{k_1! k_2! \cdots} = \sum_{n \geq 0} \sum_{r \geq 0} q^n t^r \sum_{\substack{\pi(n) \\ k(\pi)=r}} \frac{1}{k_1! k_2! \cdots}$$

$$= \prod_{j \geq 1} e^{tq^j} = \exp\left(\frac{tq}{1-q}\right)$$

$$= \sum_{r \geq 0} \frac{1}{r!}\left(\frac{tq}{1-q}\right)^r = \sum_{r \geq 0} \frac{t^r}{r!} \sum_{m \geq 0} \binom{-r}{m}(-1)^m q^{m+r}$$

$$= \sum_{r \geq 0} \frac{t^r}{r!} \sum_{n \geq 0} q^n \binom{n-1}{r-1}.$$

Equating coefficients yields

$$(22.17) \qquad \sum_{\substack{\pi(n) \\ k(\pi)=r}} \frac{k(\pi)!}{k_1! k_2! \cdots} = \binom{n-1}{r-1}.$$

If we sum over r, we get

$$(22.18) \qquad \sum_{\pi(n)} \binom{k(\pi)}{k_1, k_2, \ldots} = 2^{n-1}.$$

EXAMPLE 6. $\psi_j(q) = \exp(tq/j)$. This leads to

$$(22.19) \qquad \sum_{\substack{\pi(n) \\ k(\pi)=r}} \frac{1}{k_1! k_2! \cdots 1^{k_1} 2^{k_2} \cdots}$$

$$= \text{coefficient of } t^r \text{ in } \frac{t(t+1)\cdots(t+n-1)}{n!}.$$

Summing over r, we obtain

$$(22.2) \qquad \sum_{\pi(n)} \frac{1}{k_1! k_2! \cdots 1^{k_1} 2^{k_2} \cdots} = 1.$$

EXAMPLE 7. From Examples 1 and 2, since

$$p_E(n) = \sum_{\pi(n)} (1 + (-1)^{k(\pi)})/2,$$

$$\sum_{n \geq 0} p_E(n) q^n = \frac{1}{2} \left\{ \prod (1 - q^j)^{-1} + \prod (1 + q^j)^{-1} \right\}$$

$$= \prod (1 - q^j)^{-1} \left\{ \frac{1}{2} + \frac{1}{2} \prod \left(\frac{1 - q^j}{1 + q^j} \right) \right\}$$

$$= P(q)(1 - q + q^4 - q^9 + \cdots),$$

by (7.324). Equating coefficients, we have

(22.21) $\qquad p_E(n) = p(n) - p(n - 1^2) + p(n - 2^2) - p(n - 3^2) + \cdots .$

EXAMPLE 8. $\psi_j(q) = (1 + tq)/(1 - q) = 1 + (1 + t)q + (1 + t)q^2 + \cdots$. Then

(22.22) $\qquad \prod_{j \geq 1} \left(\frac{1 + tq^j}{1 - q^j} \right) = \sum_{n \geq 0} q^n \sum_{\pi(n)} (1 + t)^{Q(\pi)}.$

THEOREM 2. Let $L(\pi) = L(k_1, k_2, \ldots)$ be multilinear in the k_j, that is, linear in each k_j separately. Then, formally,

(22.23) $\qquad \sum_{n \geq 0} q^n \sum_{\pi(n)} L(\pi) = P(q) L \left(\frac{q}{1 - q}, \frac{q^2}{1 - q^2}, \cdots \right).$

It is sufficient to prove this in the case

$$L(k_1, k_2, \ldots) = \prod_{j \in S} k_j,$$

where S is any finite subset of the positive integers. Applying Theorem 1, we have

$$\psi_j(q) = \begin{cases} (1 - q)^{-1} & \text{for } j \notin S, \\ q + 2q^2 + 3q^3 + \cdots = q(1 - q)^{-2} & \text{for } j \in S. \end{cases}$$

Hence the generating function

$$\Phi(q) = \prod_{j \notin S} (1 - q^j)^{-1} \cdot \prod_{j \in S} \frac{q^j}{(1 - q^j)^2}$$

$$= \prod_{j \geq 1} (1 - q^j)^{-1} \cdot \prod_{j \in S} \frac{q^j}{(1 - q^j)}$$

$$= P(q) L \left(\frac{q}{1 - q}, \frac{q^2}{1 - q^2}, \cdots \right).$$

As an immediate corollary, we have the following theorem.

THEOREM 3. *If* $L(\pi) = \sum_{j \geq 1} a_j k_j$ *then*

(22.24)
$$\sum_{\pi(n)} L(\pi) = \sum_{\substack{u+v=n \\ u \geq 0, v \geq 1}} p(u) \sum_{j|v} a_j.$$

In fact, applying Theorem 2, we get

$$\Phi(q) = P(q) \sum_{j \geq 1} \frac{a_j q^j}{1 - q^j} = P(q) \sum_{v=1}^{\infty} q^v \sum_{j|v} a_j,$$

and equating coefficients leads to the required result.

EXAMPLE 9. Put $a_j = 1$ $(j \geq 1)$ in Theorem 3. Then $L(\pi) = k(\pi)$, so

(22.25)
$$\sum_{\pi(n)} k(\pi) = \sum_{\substack{u+v=n \\ u \geq 0, v \geq 1}} p(u) d(v),$$

where $d(v)$ is the number of divisors of v.

EXAMPLE 10. $a_j = j$ $(j \geq 1)$. Then $L(\pi) = \sum j k_j = n$, so

(22.26)
$$n p(n) = \sum_{\substack{u+v=n \\ u \geq 0, v \geq 1}} p(u) \sigma(v),$$

where $\sigma(v)$ is the sum of the divisors of v.

EXAMPLE 11. $a_j = \mu(j)$ $(j \geq 1)$, where μ is the Möbius function. Since

$$\sum_{j|v} \mu(j) = \begin{cases} 1 & \text{if } v = 1, \\ 0 & \text{if } v > 1, \end{cases}$$

(22.27)
$$\sum_{\pi(n)} \sum_{j \geq 1} \mu(j) k_j = p(n-1).$$

EXAMPLE 12. $a_j = 1$ $(j = 1), a_j = 0$ $(j > 1)$. Then

(22.28)
$$\sum_{\pi(n)} k_1 = \sum_{u=0}^{n-1} p(u).$$

EXAMPLE 13. $a_j = \phi(j)$ $(j \geq 1)$, where ϕ is the Euler function. Then, since $\sum_{j|v} \phi(j) = v$,

(22.29)
$$\sum_{\pi(n)} \sum_{j \geq 1} \phi(j) k_j = \sum_{\substack{u+v=n \\ u \geq 0, \ v \geq 1}} v p(u).$$

THEOREM 4. *For all* $m \geq 0$,

(22.30)
$$\sum_{\pi(n)} k_1 k_2 \cdots k_m = \sum_{\pi(n)} \binom{Q(\pi)}{m}.$$

Consider the generating function

$$\sum_{n \geq 0} q^n \sum_{m \geq 0} t^m \sum_{\pi(n)} k_1 \cdots k_m = P(q) L \left(\frac{q}{1-q}, \frac{q^2}{1-q^2}, \cdots \right),$$

where, by Theorem 2,

$$L = F(t) = A_0 + A_1 t + A_2 t^2 + \cdots$$

$$= \sum_{n \geq 0} \frac{q^{(n^2+n)/2} t^n}{(q)_n} = (-tq)_\infty = \prod_{j \geq 1} (1 + tq^j),$$

by (12.44). Thus the generating function is

$$P(q)F(t) = \prod_{j \geq 1} \left(\frac{1 + tq^j}{1 - q^j} \right) = \sum_{n \geq 0} q^n \sum_{\pi(n)} (1 + t)^{Q(\pi)},$$

by Example 8. Expanding the binomial and equating coefficients of $q^n t^m$, we obtain the required result.

THEOREM 5. *Let f be an arbitrary function of Q. Then*

$$(22.31) \qquad \sum_{\pi(n)} f(Q(\pi)) = \sum_{\pi(n)} \sum_{m \geq 0} k_1 \cdots k_m \Delta^m f(0).$$

For each fixed n, there is a polynomial f^* such that $f^*(Q) = f(Q)$ for $Q = 0, 1, 2, \ldots, n$. Then

$$f^*(Q) = \sum_{m \geq 0} \binom{Q}{m} \Delta^m f^*(0).$$

Therefore, by Theorem 4,

$$\sum_{\pi(n)} f(Q(\pi)) = \sum_{\pi(n)} f^*(Q(\pi)) = \sum_{\pi(n)} \sum_{m \geq 0} \binom{Q(\pi)}{m} \Delta^m f^*(0)$$

$$= \sum_{\pi(n)} \sum_{m \geq 0} k_1 \cdots k_m \Delta^m f^*(0)$$

$$= \sum_{\pi(n)} \sum_{m \geq 0} k_1 \cdots k_m \Delta^m f(0),$$

since $k_1 \cdots k_m = 0$ for $m > n$ and $\Delta^m f^*(0) = \Delta^m f(0)$ for $m \leq n$.

EXAMPLE 14. Let $f(Q) = Q$. Then

$$(22.32) \qquad \sum_{\pi(n)} Q(\pi) = \sum_{\pi(n)} k_1.$$

It would be interesting to find a combinatorial proof of (22.32).

EXAMPLE 15. Let $f(0) = 1$, $f(Q) = 0$ for $Q > 0$. Then for $n > 0$

$$(22.33) \qquad \sum_{\pi(n)} \{1 - k_1 + k_1 k_2 - k_1 k_2 k_3 + \cdots\} = 0.$$

This is a check formula due to Sylvester [43].

EXAMPLE 16. Let $f(1) = 1$, $f(Q) = 0$ for $Q \neq 1$. Since $Q(\pi) = 1$ if and only if $\pi = (j^{k_j})$, i.e., $n = jk_j$ for some j, $\sum_{\pi(n)} f(Q(\pi))$ counts the divisors of n. Hence

$$(22.34) \qquad d(n) = \sum_{\pi(n)} \{k_1 - 2k_1 k_2 + 3k_1 k_2 k_3 - \cdots\}.$$

THEOREM 6. *Let $f(j,k)$ be arbitrary, except that $f(j,0) = 0$. Then*

$$(22.4) \qquad \sum_{\pi(n)} \sum_{j \geq 1} f(j, k_j) = \sum_{\pi(n)} \sum_{j \geq 1} g(j) k_j,$$

where

$$(22.5) \qquad g(j) = \sum_{r|j} \mu\left(\frac{j}{r}\right) \sum_{d|r} \left(f\left(d, \frac{r}{d}\right) - f\left(d, \frac{r}{d} - 1\right)\right).$$

For each fixed $j \geq 1$, by Theorem 1,

$$\sum_{\pi(n)} q^n \sum f(j, k_j) = \prod_{i \neq j} (1 - q^i)^{-1} \sum_{k \geq 1} f(j, k) q^{jk}$$

$$= P(q)(1 - q^j) \sum_{k \geq 1} f(j, k) q^{jk}$$

$$= P(q) \sum_{k \geq 1} (f(j, k) - f(j, k - 1)) q^{jk}.$$

Now

$$q = \sum_{m \geq 1} \mu(m) \frac{q^m}{1 - q^m}, \qquad q^{jk} = \sum_{m \geq 1} \mu(m) \frac{q^{jkm}}{1 - q^{jkm}}.$$

Substituting, we get

$$\sum_{\pi(n)} q^n \sum f(j, k_j) = P(q) \sum_{k,m \geq 1} \mu(m)(f(j, k) - f(j, k - 1)) \frac{q^{jkm}}{1 - q^{jkm}}.$$

Summing over j, we obtain

$$\sum_{\pi(n)} q^n \sum_{j \geq 1} \sum f(j, k_j)$$

$$= P(q) \sum_{j,k,m \geq 1} \mu(m)(f(j, k) - f(j, k - 1)) \frac{q^{jkm}}{1 - q^{jkm}}$$

$$= P(q) \sum_{r \geq 1} \frac{q^r}{1 - q^r} \sum_{jkm = r} \mu(m)(f(j, k) - f(j, k - 1)).$$

An application of Theorem 2 yields

$$\sum_{\pi(n)} q^n \sum_{j \geq 1} \sum f(j, k_j) = \sum_{\pi(n)} q^n \sum_{j \geq 1} \sum g(j) k_j,$$

where

$$g(r) = \sum_{jkm = r} \mu(m)(f(j, k) - f(j, k - 1))$$

$$= \sum_{a|r} \mu\left(\frac{r}{a}\right) \sum_{jk = a} (f(j, k) - f(j, k - 1))$$

$$= \sum_{a|r} \mu\left(\frac{r}{a}\right) \sum_{j|a} \left(f\left(j, \frac{a}{j}\right) - f\left(j, \frac{a}{j} - 1\right)\right).$$

After a change of notation, this is the required result. An alternate form, derivable directly or by an application of Theorem 3, is

$$(22.51) \qquad \sum_{\pi(n)} \sum_{j \geq 1} f(j, k_j) = \sum_{\substack{u+v=n \\ u \geq 0, v \geq 1}} p(u) \sum_{j \mid v} \left(f\left(j, \frac{v}{j}\right) - f\left(j, \frac{v}{j} - 1\right) \right).$$

Now suppose that $f(j, k) = h(j)w(k)$ in Theorem 6, with $w(0) = 0$, and define $H(s)$ by

$$h(j) = \sum_{s \mid j} H(s), \qquad H(s) = \sum_{j \mid s} \mu\left(\frac{s}{j}\right) h(j).$$

Then we obtain

THEOREM 7. *With the above notation,*

$$(22.6) \qquad \sum_{\pi(n)} \sum_{j \geq 1} h(j)w(k_j) = \sum_{\pi(n)} \sum_{j \geq 1} k_j \sum_{r \mid j} (w(r) - w(r-1))H\left(\frac{j}{r}\right).$$

EXAMPLE 17. Let $H(s) = 1$, $h(j) = d(j)$ in Theorem 7. Then

$$(22.61) \qquad \sum_{\pi(n)} \sum_{j \geq 1} d(j)w(k_j) = \sum_{\pi(n)} \sum_{j \geq 1} k_j \sum_{r \mid j} (w(r) - w(r-1)).$$

EXAMPLE 18. Let $w(k) = 1$ for $k > 0$ in Theorem 7. Then

$$(22.62) \qquad \sum_{\pi(n)} \sum_{\substack{j \geq 1 \\ k_j > 0}} h(j) = \sum_{\pi(n)} \sum_{j \geq 1} H(j)k_j = \sum_{\substack{u+v=n \\ u \geq 0, v \geq 1}} p(u)h(v).$$

EXAMPLE 19. In Example 18, let $h(j) = d(j)$, so that $H(s) = 1$. Then

$$(22.63) \qquad \sum_{\pi(n)} \sum_{\substack{j \geq 1 \\ k_j > 0}} d(j) = \sum_{\pi(n)} k(\pi) = \sum_{\substack{u+v=n \\ u \geq 0, v \geq 1}} p(u)d(v).$$

EXAMPLE 20. Let $H(s) = s$, $h(j) = \sigma(j)$ in Example 18. Then

$$(22.64) \qquad \sum_{\pi(n)} \sum_{\substack{j \geq 1 \\ k_j > 0}} \sigma(j) = np(n) = \sum_{\substack{u+v=n \\ u \geq 0, v \geq 1}} p(u)\sigma(v).$$

In Theorem 7, put $H(s) = \delta(s, 1)$ (Kronecker delta). Then $h(j) = 1$, and we have

THEOREM 8. *If $w(0) = 0$, then*

$$(22.7) \qquad \sum_{\pi(n)} \sum_{j \geq 1} w(k_j) = \sum_{\pi(n)} \sum_{j \geq 1} (w(j) - w(j-1))k_j.$$

EXAMPLE 21. Let $\delta_i(\pi) = $ the number of j such that $k_j(\pi) = i$ $(i \geq 1)$. Then putting $w(k) = 1$ for $k = i$, $w(k) = 0$ for $k \neq i$ in Theorem 8, we have

$$(22.71) \qquad \sum_{\pi(n)} \delta_i(\pi) = \sum_{\pi(n)} (k_i - k_{i+1}).$$

EXAMPLE 22. Let $w(k) = 0$ for $k < i$, $w(k) = 1$ for $k \geq i$ $(i > 0)$. Then

$$(22.72) \qquad \sum_{\substack{\pi(n) \ j \geq 1 \\ k_j \geq i}} \sum 1 = \sum_{\pi(n)} k_i.$$

That is, the total number of frequencies $\geq i$ in all partitions of n is equal to the total number of times that the part i occurs. This generalizes Example 14.

EXAMPLE 23. Let $w(k) = k(k+1)/2$. Then

$$(22.73) \qquad \sum_{\pi(n)} \sum_{j \geq 1} k_j(k_j + 1)/2 = np(n).$$

In Theorem 7, take $h(j) = \delta(j, j_0)$ (Kronecker delta) and let $w(k) = \lambda([k/m])$, where m is a positive integer and $[\,]$ denotes the greatest integer function. Then we obtain, after dropping the subscript in j_0,

EXAMPLE 24. $\sum_{\pi(n)} \lambda([k_j/m]) = \sum_{\pi(n)} \lambda(k_{jm})$.
This holds even when $\lambda(0) \neq 0$. It may be proved without the use of generating functions or the preceding theorems by setting up a bijection $\pi \to \pi'$ of the partitions of n such that

$$(22.74) \qquad \left[\frac{k_j(\pi)}{m} \right] = k_{jm}(\pi').$$

23. Partitions with odd parts and with distinct parts. In (7.5), put $b = i$. Then

$$HS = \frac{1}{1+i} \prod_{n \geq 1} \frac{1}{1+q^{2n}} \sum_{m \geq 0} q^{(m^2+m)/2},$$

$$(23.1) \qquad G = -\frac{1}{1-i} \sum_{n \geq 0} \frac{q^n}{(-q^2; q^2)_n} = -\frac{1}{1-i} F(0, -1; q : q^2),$$

$$F = (1-i)F(i, 0; i) = \sum_{n \geq 0} i^n q^{(3n^2+n)/2} - i \sum_{n \geq 1} i^n q^{(3n^2-n)/2},$$

the last by reason of (7.7). Multiply $HS = F + G$ by $(1 - i)$, equate real parts, and replace q by $-q$ to get

(23.11)

$F(0, -1; -q : q^2)$

$$= 1 + \sum_{n \geq 1} \{(q^{n(6n+1)} - q^{n(6n-1)}) + (q^{(2n-1)(3n-1)} - q^{(2n-1)(3n-2)})\}$$

$$= 1 + \sum_{n \geq 1} (q^{(3n^2+n)/2} - q^{(3n^2-n)/2}) = 2F(-1, 0; -1),$$

again by (7.7). Now by (6.3), followed by (2.4),
(23.12)
$$F(0, -1; -q : q^2) = \frac{2}{1+q} F(0, -q; -1 : q^2)$$

$$= \frac{1}{1+q} \{1 + q - qF(0, -q; -q^2 : q^2)\}$$

$$= 1 - \frac{q}{1+q} \sum_{n \geq 0} \frac{(-1)^n q^{2n}}{(-q^3; q^2)_n} = 1 - \frac{q}{1+q} + \frac{q^3}{(1+q)(1+q^3)}$$

$$- \frac{q^5}{(1+q)(1+q^3)(1+q^5)} + \cdots .$$

Also by (2.4),

(23.13)
$$2F(-1, 0; -1) = 1 - qF(-1, 0; -q)$$
$$= 1 - q + (1+q)q^2 - (1+q)(1+q^2)q^3 + \cdots .$$

Combining (23.11), (23.12), and (23.13), we have
(23.2)
$$1 - q + (1+q)q^2 - \cdots + (-1)^n (1+q) \cdots (1+q^{n-1})q^n + \cdots = 1 - \frac{q}{1+q}$$

$$+ \frac{q^3}{(1+q)(1+q^3)} - \cdots + (-1)^n \frac{q^{2n-1}}{(1+q)(1+q^3) \cdots (1+q^{2n-1})} + \cdots$$

$$= 1 - q + q^2 - q^5 + q^7 - \cdots .$$

Now define

(23.3)
$$Q_a(n) = \text{the number of partitions of } n \text{ into distinct parts,}$$
$$\text{the maximum part being } \equiv a \ (\text{mod } 2), \ a = 0, 1;$$
$$Q_b^*(n) = \text{the number of partitions of } n \text{ into odd parts, the}$$
$$\text{maximum part being } \equiv b \ (\text{mod } 4), \ b = 1, 3.$$

The well-known identity

(23.4)
$$\prod_{n \geq 1} \frac{1}{1 - q^{2n-1}} = \prod_{n \geq 1} (1 + q^n)$$

can obviously be paraphrased into the theorem which in our notation is

(23.5)
$$Q_0(n) + Q_1(n) = Q_1^*(n) + Q_3^*(n).$$

Now the first series in (23.2) is clearly the generating function for $Q_0(n) - Q_1(n)$. Examination will reveal that the second series generates $(-1)^n (Q_1^*(n) - Q_3^*(n))$. Hence

(23.6)
$$Q_0(n) - Q_1(n) = (-1)^n (Q_1^*(n) - Q_3^*(n)).$$

From the third series in (23.2) we obtain this common value explicitly:

(23.7)
$$Q_0(n) - Q_1(n) = \begin{cases} +1 & \text{if } n = \frac{3k^2+k}{2}, \ k \geq 0, \\ -1 & \text{if } n = \frac{3k^2-k}{2}, \ k > 0, \\ 0 & \text{otherwise.} \end{cases}$$

It follows readily from (23.5) and (23.6) that

(23.8)
$$Q_1^*(2n) = Q_0(2n); \qquad Q_3^*(2n) = Q_1(2n),$$
$$Q_1^*(2n+1) = Q_1(2n+1); \qquad Q_3^*(2n+1) = Q_0(2n+1).$$

The results (23.2), (23.7), and (23.8) were stated in [16]. Afterwards, D. H. Lehmer communicated to me elegant graphical proofs of (23.7) and (23.8).

Another theorem along these lines can be obtained from (20.72). In that identity, replace a by tq, multiply through by tq, and apply definition (1.1) to get

(23.9)
$$\frac{qt}{1-qt} + \frac{q^3t^2}{(1-qt)(1-q^3t)} + \frac{q^5t^3}{(1-qt)(1-q^3t)(1-q^5t)} + \cdots$$
$$= qt + (1+q)q^2t^2 + (1+q)(1+q^2)q^3t^3 + \cdots .$$

By equating coefficients of $q^N t^M$, we find:

(23.91) *The number of partitions of N into distinct parts with maximum part M is equal to the number of partitions of N into odd parts such that the maximum part plus twice the number of parts is $2M + 1$.*

It is not difficult to derive (23.8) from (23.91), by summing separately over even and odd M, and by taking into account the fact that the number of parts has the same parity as N when all the parts are odd.

24. Continuation. From (8.4) and (8.41), with b replaced by aq, we obtain

(24.1)
$$F(-a^{-1}q^{-1}, 0; aq) = \sum_{k \geq 0} a^k q^k V_k.$$

The transformation (2.2) yields

(24.11)
$$1 + (1+a)qF(-a^{-1}, 0; aq) = \sum_{k \geq 0} a^k q^k V_k.$$

The coefficient of a^k on both sides of (24.11) can be interpreted arithmetically. Indeed, by (1.1),

(24.2)
$$qF(-a^{-1}, 0; aq) = q + (1+a^{-1}q)aq^2 + \cdots$$
$$+ (1+a^{-1}q)\cdots(1+a^{-1}q^n)a^n q^{n+1} + \cdots .$$

A typical term on the right of (24.2) is $a^k q^N = a^{-t}a^n q^N$, where

$$N = n+1+p_1 + p_2 + \cdots + p_t, \qquad n+1 > p_1 > p_2 > \cdots > p_t > 0.$$

For t, n, and N fixed we obtain all the partitions of N into $t+1$ distinct parts of which $n+1$ is the largest; thus $k = (n+1) - (t+1)$ is the difference between the largest part and the number of parts. We are therefore led to the definition, due to Dyson:

(24.3) *The rank of a partition is the difference between the largest part and the number of parts.*

For arbitrary partitions the rank may be positive, negative, or zero; for partitions into distinct parts the rank cannot be negative. Denoting by $\delta_k(N)$ the number of those partitions of N into distinct parts which have rank k, we see that the coefficient of a^k in (24.2) is

$$\sum_{N\geq 1} \delta_k(N)q^N.$$

Thus, (24.11) becomes

$$(24.31) \qquad 1 + \sum_{k\geq 0} a^k \sum_{N\geq 1} (\delta_k(N) + \delta_{k-1}(N))q^N = \sum_{k\geq 0} a^k q^k V_k,$$

where $\delta_{-1}(N) = 0$. Hence

$$(24.4) \qquad \sum_{N\geq 1} (\delta_k(N) + \delta_{k-1}(N))q^N = q^k V_k \qquad (k > 0).$$

Now for k odd, (8.5b) shows that

$$(24.5) \qquad q^k V_k = \sum_{N\geq 1} \omega_k(N)q^N,$$

where $\omega_k(N)$ is the number of partitions of N into odd parts with maximum part equal to k. Equating coefficients in (24.4) and (24.5), and writing $k = 2r + 1$, $r \geq 0$, we have the result stated in [16]:

$$(24.6) \qquad \omega_{2r+1}(N) = \delta_{2r+1}(N) + \delta_{2r}(N) \qquad (r \geq 0, \ N \geq 1).$$

This theorem gives us a refined correspondence between partitions into odd and distinct parts.

It is possible to derive an arithmetic theorem for even values of k in (24.4), but the result is not very elegant.

25. The rank of a partition. In this section we shall study more closely the concept of rank (24.3). We define $P_r(n)$ as the number of partitions of n with rank r, and $P_r(n; Q)$ as the number of partitions of n with rank $\equiv r \pmod{Q}$. We make the convention that $P_0(0) = 1$, $P_r(n) = 0$ for $r \neq 0$, $n \leq 0$ and $r = 0$, $n < 0$. Let

$$(25.1) \qquad K_r(q) = \sum_n P_r(n)q^n,$$

$$(-\infty < r < \infty)$$

$$(25.11) \qquad K_r(q; Q) = \sum_n P_r(n; Q)q^n.$$

Our first task will be to determine these generating functions explicitly.

For $|q| < |t| < |q|^{-1}$, we may expand the series

$$(25.12) \qquad \sum_{m\geq 1} \frac{t^{m-1}q^m}{(1 - t^{-1}q)(1 - t^{-1}q^2) \cdots (1 - t^{-1}q^m)}$$

in powers of t and q as follows

$$\sum_{m\geq 1} t^{m-1}q^m \sum_{p_1 \geq 0} \cdots \sum_{p_m \geq 0} t^{-(p_1 + \cdots + p_m)}q^{p_1 + 2p_2 + \cdots + mp_m}.$$

The coefficient of $t^r q^n$ is then the number of solutions of

$$n = p_1 + 2p_2 + \cdots + (p_m + 1)m \qquad (p_i \geq 0),$$
$$r = m - (p_1 + p_2 + \cdots + (p_m + 1)).$$

Each solution corresponds to a partition of n into $\nu = p_1 + p_2 + \cdots + (p_m + 1)$ parts with maximum part m, so that $r = m - \nu$ is the rank of the partition. Hence (25.12) may be written

$$(25.13) \qquad \sum_{r=-\infty}^{+\infty} \sum_{n \geq 1} P_r(n) t^r q^n = \sum_{r=-\infty}^{+\infty} t^r K_r(q) - 1.$$

But, directly from definition, we have (25.12) equal to $t^{-1} F(0, t^{-1}; tq) - t^{-1}$. Thus, by (2.4),

$$(25.14) \qquad (1 - t) F(0, t^{-1}; t) = \sum_{r=-\infty}^{+\infty} K_r(q) t^r \qquad (|q| < |t| < |q|^{-1}).$$

Now, by (16.4),

$$(25.15) \quad (1 - t) F(0, t^{-1}; t)(q)_\infty = 1 + (1 - t)(1 - t^{-1}) \sum_{n \geq 1} \frac{C_n}{(1 - tq^n)(1 - t^{-1}q^n)},$$

where

$$(25.16) \qquad C_n = (-1)^n (1 + q^n) q^{(3n^2 + n)/2}.$$

After the denominators are expanded, the right side of (25.15) becomes

$$1 + (1 - t)(1 - t^{-1}) \sum_{N=-\infty}^{+\infty} t^N W_N,$$

where

$$W_N = \sum_{n \geq 1} C_n \sum_{\substack{r-s=N \\ r,s \geq 0}} q^{n(r+s)}.$$

Clearly $W_{-N} = W_N$, and for $N \geq 0$,

$$W_N = \sum_{n \geq 1} \sum_{s \geq 0} C_n q^{n(2s+N)} = \sum_{n \geq 1} (-1)^n \frac{q^{(3n^2+n)/2+Nn}}{1 - q^n}.$$

Hence

$$(25.17) \qquad (1 - t) F(0, t^{-1}; t)(q)_\infty = 1 + \sum_{r=-\infty}^{+\infty} t^r (2W_r - W_{r-1} - W_{r+1}).$$

Comparing (25.17) with (25.14), we have

$$(25.18) \qquad K_0(q) = \frac{1}{(q)_\infty} \left\{ 1 + 2 \sum_{n \geq 1} (-1)^n q^{(3n^2+n)/2} \right\},$$

(25.19) $K_{-r}(q) = K_r(q) = \dfrac{1}{(q)_\infty} \sum_{n\geq 1} (-1)^{n+1}(1-q^n)q^{(3n^2-n)/2+rn}$ $(r > 0)$.

Now, by summing over all r in a residue class $(\bmod\, Q)$, we find easily that

(25.20) $K_0(q; Q) = \dfrac{1}{(q)_\infty}\left\{ 1 + 2\sum_{n\geq 1}(-1)^n\dfrac{1-q^{n(Q-1)}}{1-q^{nQ}}q^{(3n^2+n)/2}\right\}$,

(25.21) $K_r(q; Q) = \dfrac{1}{(q)_\infty}\sum_{n\geq 1}(-1)^{n+1}\left(\dfrac{1-q^n}{1-q^{nQ}}\right)(q^{rn}+q^{(Q-r)n})q^{(3n^2-n)/2}$,

for $0 < r < Q$. The generating functions (25.19) and (25.21) were given by Dyson in [**13**], along with a number of interesting conjectures, which have since been proved by Atkin and Swinnerton-Dyer [**3**].

In this section, we find it necessary to refer ahead to §26.

The function $G(t) = (1-t)F(0, t^{-1}; t)$ is obviously of importance in the study of the generating functions $K_r(q)$ and $K_r(q; Q)$. Since some of Ramanujan's mock theta-functions are special cases of $G(t)$, we may expect that identities involving them can be translated into arithmetic theorems. For example, we have

(25.3) $f(q) = 2F(0, -1; -1) = K_0(q; 2) - K_1(q; 2)$,

(25.31) $\phi(q) = (1 - i)F(0, -i; i) = K_0(q; 4) - K_2(q; 4)$,

(25.32)
$\chi(q) = (1 + \omega^2)F(0, -\omega; -\omega^2)$
$= K_0(q; 6) + K_1(q; 6) - K_2(q; 6) - K_3(q; 6)$,

where $\omega = \exp(2\pi i/3)$. Now, by (26.56), with q replaced by $-q$, we have

(25.33) $\phi(-q) = 1 - q + (1+q)q^3 - (1+q)(1+q^3)q^5 + \cdots$.

The coefficient of q^n is easily seen to be the excess of the number of partitions of n into distinct odd parts with maximum part of the form $4k + 3$ over the number of such partitions with maximum part of the form $4k + 1$. Denoting this arithmetic function by $\Delta(n)$, and comparing coefficients with those in (25.31), we find

(25.34) $P_0(n; 4) - P_1(n; 4) = (-1)^n\Delta(n)$.

In §26, we shall find a simple interpretation for $\psi(q)$ (26.13) as the generating function for $\beta(n)$, the number of partitions of n into odd parts without gaps. By (26.66), we have

(25.35) $2\psi(q) = \phi(q) - f(-q)$,

that is,

$2\sum_{n\geq 1}\beta(n)q^n = K_0(q; 4) - K_2(q; 4) - K_0(-q; 2) + K_1(-q; 2)$

$= \sum_{n\geq 1}q^n(P_0(n; 4) - P_2(n; 4) - (-1)^n P_0(n; 2) + (-1)^n P_1(n; 2))$.

Taking the exponent even, we get

$$2\beta(2n) = P_0(2n;4) - P_2(2n;4) - P_0(2n;4) - P_2(2n;4) + P_1(2n;4) + P_3(2n;4)$$
$$= 2(P_1(2n;4) - P_2(2n;4)),$$

and similarly for an odd exponent, yielding finally

(25.36) $$\beta(2n) = P_1(2n;4) - P_2(2n;4),$$

(25.37) $$\beta(2n+1) = P_0(2n+1;4) - P_1(2n+1;4).$$

It may be of interest to compare these results with the conjecture of Dyson [13], since proved in [3], that for $n \equiv 4 \pmod 5$,

$$P_0(n;5) = P_1(n;5) = P_2(n;5) = P_3(n;5) = P_4(n;5).$$

Equations (25.36) and (25.37) show that the four classes of partitions (mod 4) are *never* equinumerous, since $\beta(n)$ is always positive.

We adopt the convention that an arithmetical function defined only for non-negative or positive integers shall take the value 0 for all other arguments. Thus $p(-4) = 0$, $E_1(n/2;6) = 0$ unless n is a positive even integer, and so forth.

A similar translation of the identity (26.72) yields the result

(25.38) $$2P_2(n;6) + P_3(n;6) = 2\sum_{k \geq 1}(-1)^k E_1(k;4)p(n-3k).$$

From an identity of Watson [45],

(25.4) $$2\rho(q) + \omega(q) = 3\prod_{n \geq 1}\frac{(1+q^{3n})^4(1-q^{3n})^2}{(1-q^{2n})},$$

we can also obtain a paraphrase. Defining

$$H_r(k) = P_r(k) + P_{r-1}(k) + \cdots,$$

that is, the number of partitions of k with rank not exceeding r, we have, for all $N \geq 0$,

(25.41) $$\sum_{3r+2k=N}(H_{3r+1}(k) - p(k)E_1(4r+1;4)) = 0,$$

where, of course, $E_1(m;4) = 0$ for $m < 0$.

The relations (26.88) yield the following arithmetic results:

(25.5) $$H_{r+1}(n) = H_{-r+1}(n+r) \qquad (2n+r \geq 0),$$

(25.51) $$\sum_{r+n=2N+1}(-1)^r H_{2r+1}(n) + \sum_{r+2n=N}H_{r+1}(n) = 0 \qquad (N \geq 0),$$

(25.52) $$\sum_{r+n=2N}(-1)^r H_{2r+1}(n) = \sum_{n+2r=N}E_1(4n+1;4)p(r) \qquad (N \geq 0),$$

(25.53) $$\sum_{r+n=N}(-1)^r H_{2r}(n) = 0 \qquad (N \geq 1).$$

By using the transformations $b \to bq^{-1}$, $t \to tq$, it is not difficult to show that the function $G(t) = (1 - t)F(0, t^{-1}; t)$ satisfies

(25.6) $$qt^3(1 - qt)G(t) + (1 - t)G(qt) = (1 - t)(1 - qt)(1 - qt^2).$$

Using (25.14) and equating coefficients of t^r, $r > 4$, we obtain

(25.61) $$q^{r-1}K_r - q^{r-2}K_{r-1} + K_{r-3} - qK_{r-4} = 0 \qquad (r > 4).$$

The five relations for $0 \le r \le 4$ are not independent; the essential ones are

(25.62) $$qK_0 - K_1 + q^2K_3 - q^3K_4 = q,$$

(25.63) $$K_0 - qK_1 - qK_2 + q^2K_3 = 1 + q.$$

Equating coefficients for $r < 0$ yields nothing new. The arithmetic equivalents of (25.61), (25.62), and (25.63) are easy to write down:

(25.611) $P_{r+1}(n) - P_r(n-1) = P_{r+3}(n-r-2) - P_{r+4}(n-r-3) \qquad (r > 4)$,

(25.621) $$P_0(n-1) - P_1(n) + P_3(n-2) - P_4(n-3) = 0 \qquad (n > 1),$$

(25.631) $$P_0(n) - P_1(n-1) - P_2(n-1) + P_3(n-2) = 0 \qquad (n > 1).$$

If we sum (25.61) for $r \ge 5$, we get

$$qK_1 + q^3K_4 = (1 - q)\sum_{r \ge 2} K_r = \frac{1}{2}(1 - q)((q)_\infty^{-1} - K_0 - 2K_1),$$

that is

(25.7) $$(1 - q)(q)_\infty^{-1} = (1 - q)K_0 + 2K_1 + 2q^3K_4.$$

Eliminating $K_1 + q^3K_4$ by (25.62), we obtain

(25.71) $$(1 - q)(q)_\infty^{-1} = -2q + (1 + q)K_0 + 2q^2K_3,$$

which has the paraphrase

(25.72) $$p(n + 1) - p(n) = P_0(n + 1) + P_0(n) + 2P_3(n - 1) \qquad (n \ge 1).$$

Thus the function $p(n)$ is completely determined by the two functions $P_0(n)$, $P_3(n)$. In fact,

(25.73) $$p(N + 1) = P_0(N + 1) + 2P_0(N) + 2\sum_{n=1}^{N-1}(P_0(n) + P_3(n)) \qquad (N \ge 1).$$

It is convenient at this point to introduce the summatory function

(25.8) $$S_R(n) = \sum_{|r| \le R} P_r(n) \qquad (R \ge 0).$$

From (25.18) and (25.19) it is easy to see that

(25.81) $$\sum_n S_R(n)q^n = (q)_\infty^{-1}\left\{1 + 2\sum_{m \ge 1}(-1)^m q^{(3m^2+m)/2+Rm}\right\}.$$

Hence

(25.82) $S_R(n) = p(n) + 2 \sum_{m \geq 1} (-1)^m p\left(n - \frac{3m^2 + (2R+1)m}{2}\right).$

If $R > \frac{1}{2}(n - 7)$, we need only the term for $m = 1$, so

(25.821) $S_R(n) = p(n) - 2p(n - R - 2)$ $(R > \frac{1}{2}(n - 7)).$

Replacing R by $R - 1$ and subtracting, we obtain

$P_R(n) = p(n - R - 1) - p(n - R - 2)$ $(R > \frac{1}{2}(n - 5)).$

We write this in the somewhat more convenient form

(25.83) $P_R(k + R) = p(k - 1) - p(k - 2)$ $(R > 0,\ R > k - 5).$

It should be possible to prove this directly from the definition by a graph-theoretic argument.

We shall now develop a new form for $K_r(q)$, which in turn will lead us to a new identity. Every partition of N with rank $r \geq 0$ and maximum part n has, in its graphical representation, an outside shell of $2n - r - 1 = n + (n - r) - 1$ nodes, and a residual content $s = N - (2n - r - 1)$, which may be partitioned subject to the restrictions that the maximum part does not exceed $n - 1$ and the number of parts does not exceed $n - r - 1$. The number of such partitions is the coefficient of q^s in

$$\begin{bmatrix} (n - r - 1) + (n - 1) \\ n - 1 \end{bmatrix} = \begin{bmatrix} 2n - r - 2 \\ n - 1 \end{bmatrix},$$

[1; Chapter 3], or the coefficient of q^N in

$$\begin{bmatrix} 2n - r - 2 \\ n - 1 \end{bmatrix} q^{2n-r-1}.$$

Summing for all $n > r$, we obtain (for $r > 0$)

$$K_r(q) = \sum_{n > r} \begin{bmatrix} 2n - r - 2 \\ n - 1 \end{bmatrix} q^{2n-r-1},$$

or, replacing n by $m + r + 1$, $m \geq 0$,

(25.9) $K_r(q) = \sum_{m \geq 0} \begin{bmatrix} 2m + r \\ m \end{bmatrix} q^{2m+r+1}$ $(r > 0).$

For $r = 0$ we must add the single partition of 0, of rank 0, so that

(25.91) $K_0(q) = 1 + \sum_{m \geq 0} \begin{bmatrix} 2m \\ m \end{bmatrix} q^{2m+1}.$

Now we have

$$\begin{bmatrix} 2m + r \\ m \end{bmatrix} = \frac{(q)_{2m+r}}{(q)_m (q)_{m+r}} = \frac{(q^{r+1})_{2m}}{(q)_m (q^{r+1})_m}.$$

Thus we are led to consider the function

(25.92) $$Q(a;t) = \sum_{m \geq 0} \frac{(aq)_{2m}}{[m][m;a]} t^m,$$

which reduces to $K_r(q)$ (essentially) for $a = q^r$, $t = q^2$. In fact

(25.93) $$\begin{aligned} K_r(q) &= q^{r+1}Q(q^r;q^2) \qquad (r > 0), \\ K_0(q) &= 1 + qQ(1;q^2). \end{aligned}$$

But

$$\begin{aligned} Q(a;t) &= \sum_{m \geq 0} \frac{(aq^{m+1})_m}{(q)_m} t^m \\ &= \sum_{m \geq 0} \frac{t^m}{(q)_m} \sum_{k=0}^{m} \begin{bmatrix} m \\ k \end{bmatrix} (-aq^m)^k q^{(k^2+k)/2} \\ &= \sum_{k \geq 0} (-a)^k q^{(k^2+k)/2} \sum_{m \geq k} \begin{bmatrix} m \\ k \end{bmatrix} \frac{(q^k t)^m}{(q)_m} \\ &= \sum_{k \geq 0} (-a)^k q^{(k^2+k)/2} \sum_{n \geq 0} \begin{bmatrix} n+k \\ k \end{bmatrix} \frac{(q^k t)^{n+k}}{(q)_{n+k}} \\ &= \sum_{k \geq 0} \frac{(-at)^k}{(q)_k} q^{(3k^2+k)/2} \sum_{n \geq 0} \frac{(q^k t)^n}{(q)_n} \\ &= \sum_{k \geq 0} \frac{(-at)^k}{(q)_k} q^{(3k^2+k)/2} (tq^k)_\infty^{-1}. \end{aligned}$$

Thus we have proved the identity

(25.94) $$Q(a;t) = (t)_\infty^{-1} \sum_{k \geq 0} \frac{(t)_k}{(q)_k} (-at)^k q^{(3k^2+k)/2}.$$

In view of (25.93), this affords us an alternative proof of (25.18) and (25.19). There are many other interesting special cases, of which we give one example. Set $a = -1$, $t = q$. Then

(25.95) $$\begin{aligned} &1 + \frac{(1+q^2)}{(1-q)}q + \frac{(1+q^3)(1+q^4)}{(1-q)(1-q^2)}q^2 + \frac{(1+q^4)(1+q^5)(1+q^6)}{(1-q)(1-q^2)(1-q^3)}q^3 + \cdots \\ &= (q)_\infty^{-1} \sum_{k \geq 0} q^{3(k^2+k)/2} = \prod_{n \geq 1} \frac{(1+q^{3n})(1-q^{6n})}{(1-q^n)}. \end{aligned}$$

In a manner similar to the proof of (25.94), we can prove the more general identity

(25.96) $$\sum_{m \geq 0} \frac{(aq)_{2m}(bq)_m}{(aq)_m(q)_m} t^m = \frac{(btq)_\infty}{(t)_\infty} \sum_{k \geq 0} \frac{(bq)_k(t)_k}{(q)_k(btq)_{2k}} (-at)^k q^{(3k^2+k)/2}.$$

Notes

In 1948, the author published a short note [10] describing several of the partition theorems that were derivable from the work presented in Chapter 1. This chapter contains all of the results announced in 1948 and many others as well. An extensive account of the theory of partitions is given in [4] (cf. [7, Chapters 6 and 7]).

§22. This chapter begins with a detailed account of Fine's paper "Sums over Partitions," which has previously only appeared in a conference proceedings (Report of the Institute of the Theory of Numbers, Boulder, 1959) and has therefore not been widely available. The methods developed in this section were given wide publicity by J. Riordan in his book *Combinatorial Identities* (John Wiley, New York, 1968). Riordan devotes Section 5.5 (pp. 182–187 and p. 199) to Fine's paper.

§23. The results of this section have been discussed in [1], [2], and [3]. Extensions connecting these results with those of L. J. Rogers are found in [5].

§24. The rank of a partition was introduced by F. J. Dyson [9]. In that paper Dyson also conjectured the existence of the crank of a partition. This was to be a statistic of partitions that would provide a combinatorial interpretation of Ramanujan's congruence $p(11n + 6) \equiv 0 \pmod{11}$. In [11], F. Garvan found such a statistic for certain vector partitions, and in [8] the crank for ordinary partitions was presented. Identity (24.6), announced in [10], has been proved via a different derivation in [2]. A purely arithmetic-combinatorial proof appears in [6].

References

1. G. E. Andrews, *On basic hypergeometric series, mock theta functions and partitions.* I, Quart. J. Math. Oxford Ser. (2) **17** (1966), 64–80.

2. ____, *On basic hypergeometric series, mock theta functions and partitions.* II, Quart. J. Math. Oxford. Ser. (2) **17** (1966), 132–143.

3. ____, *q-Identities of Auluck, Carlitz and Rogers,* Duke Math. J. **33** (1966), 575–582.

4. ____, *The theory of partitions,* Encyclopedia of Mathematics and its Applications, Vol. 2, G.-C. Rota, editor, Addison-Wesley, Reading, Mass., 1976; reprinted by Cambridge Univ. Press, 1984.

5. ____, *Partitions: Yesterday and today,* New Zealand Math. Soc., Wellington, 1979.

6. ____, *On a partition theorem of N. J. Fine,* J. Nat. Acad. Math. India **1** (1983), 105–107.

7. ____, *q-Series: Their development and application in analysis, number theory, combinatorics, physics and computer algebra,"* CBMS Regional Conf. Ser. in Math., no. 66, Amer. Math. Soc., Providence, R.I., 1986.

8. G. E. Andrews and F. G. Garvan, *Dyson's crank of a partition,* Bull. Amer. Math. Soc. (to appear).

9. F. J. Dyson, *Some guesses in the theory of partitions,* Eureka, Feb., 1944, 10–15.

10. N. J. Fine, *Some new results on partitions,* Proc. Nat. Acad. Sci. U.S.A. **34** (1948), 616–618.

11. F. G. Garvan, *New combinatorial interpretations of Ramanujan's partition congruences mod 5, 7 and 11,* Trans. Amer. Math. Soc. **305** (1988), 47–77.

CHAPTER 3

Mock theta functions and the functions $J(N)$, $L(N)$

26. Mock theta functions. In §25 we showed that the function $F(0, t^{-1}; t)$ is of significance in the study of the rank of partitions. In this section we shall connect the same function with the mock theta functions introduced by Ramanujan [23; p. 354] and later discussed by Watson [45], Dragonette [12], Andrews and Agarwal (see Chapter Notes). All those of third order are special cases of $F(0, t^{-1}; t)$; thus the methods developed here enable us to obtain the relations among them given by the first two authors mentioned above, and to find other forms of them.

Ramanujan's list consists of the following four:

$$(26.11) \quad f(q) = \sum_{n \geq 0} \frac{q^{n^2}}{(-q)_n^2},$$

$$(26.12) \quad \phi(q) = 1 + \frac{q}{1+q^2} + \frac{q^4}{(1+q^2)(1+q^4)} + \frac{q^9}{(1+q^2)(1+q^4)(1+q^6)} + \cdots,$$

$$(26.13) \quad \psi(q) = \frac{q}{1-q} + \frac{q^4}{(1-q)(1-q^3)} + \frac{q^9}{(1-q)(1-q^3)(1-q^5)} + \cdots,$$

$$(26.14) \quad \chi(q) = 1 + \frac{q}{1-q+q^2} + \frac{q^4}{(1-q+q^2)(1-q^2+q^4)} + \cdots.$$

First, putting $\theta = \pi$ in (12.32), we find that

$$(26.21) \qquad\qquad f(q) = 2F(0, -1; -1),$$

and by (2.4) and definition (1.1),

$$(26.22) \qquad f(q) = 2 - F(0, -1; -q) = 1 + \sum_{n \geq 1} \frac{(-1)^{n-1} q^n}{(-q)_n}.$$

Next, setting $t = -1$ in (11.4), we have

$$(26.23) \qquad \frac{(-q)_\infty^2}{(q)_\infty} f(q) = 1 + 4 \sum_{n \geq 1} \frac{(-q)_{n-1}^2}{(q)_n} q^n.$$

Again, with $b = t = -1$ in (15.51),

$$(26.24) \qquad (-q)_\infty f(q) = 2 \sum_{n \geq 0} \frac{q^{(n^2+n)/2}}{(q)_n (1+q^n)}.$$

Equation (26.22) yields an arithmetic interpretation of $f(q)$. For we may write

$$f(q) = 1 + \frac{1}{(-q)_\infty} \sum_{k \geq 1} (-1)^{k-1} q^k (1 + q^{k+1})(1 + q^{k+2}) \cdots ,$$

and $q^k (1 + q^{k+1})(1 + q^{k+2}) \cdots$ generates the number of partitions of n into distinct parts with minimum part k. Denoting this function by $L_k(n)$, we have

$$(-q)_\infty f(q) = (-q)_\infty + \sum_{n \geq 1} (L_1(n) - L_2(n) + L_3(n) - \cdots)q^n$$

$$= 1 + 2 \sum_{n \geq 1} (L_1(n) + L_3(n) + L_5(n) + \cdots)q^n.$$

Hence, defining

(26.25) $$L(n) = L_1(n) + L_3(n) + L_5(n) + \cdots ,$$

we have

(26.26) $$(-q)_\infty f(q) = 1 + 2 \sum_{n \geq 1} L(n)q^n.$$

Now, setting $t = -1$ in (25.15), we obtain

(26.27)
$$f(q) = (q)_\infty^{-1} \left\{ 1 + 4 \sum_{n \geq 1} (-1)^n \frac{q^{(3n^2+n)/2}}{1 + q^n} \right\}$$

$$= (q)_\infty^{-1} \left\{ 1 + 4 \sum_{N \geq 1} J(N)q^N \right\}.$$

Comparing (26.26) and (26.27), we see that the functions $L(N)$ and $J(N)$ are related by

(26.28)
$$\left\{ 1 + 2 \sum_{N \geq 1} L(N)q^N \right\} \left\{ 1 + 2 \sum_{m \geq 1} (-1)^m q^{m^2} \right\}$$

$$= 1 + 4 \sum_{N \geq 1} J(N)q^N,$$

where we have used (7.324) for $(q)_\infty/(-q)_\infty$. Hence, if we write $\varepsilon(N) = (-1)^N$ if N is a square, $= 0$ otherwise, equating coefficients in (26.28) yields, for $N \geq 1$,

(26.29) $$2J(N) = L(N) + \varepsilon(N) - 2 \sum_{1 \leq m < \sqrt{N}} (-1)^{m-1} L(N - m^2).$$

An immediate consequence is the result that $L(N)$ is odd if and only if N is a square. This could also be deduced from (12.45), which implies that $L(N) \equiv d(N) \pmod 2$. It is clear, however, that much more information can be derived from (26.29), since the function $J(N)$ is amenable to further study. We shall reserve this for §27.

Returning to the other mock theta functions, we see immediately from (12.32), with $\theta = \pi/2$, that

$$(26.31) \qquad\qquad \phi(q) = (1 - i)F(0, -i; i).$$

By (11.4), with $t = i$, we have

$$(26.32) \qquad \phi(q) = \prod_{n \geq 1} \left(\frac{1 - q^n}{1 + q^{2n}}\right) \left\{1 + 2\left(\frac{q}{(q)_1} + (1 + q^2)\frac{q^2}{(q)_2}\right.\right.$$

$$\left.\left. + (1 + q^2)(1 + q^4)\frac{q^3}{(q)_3} + \cdots\right)\right\}.$$

By (15.51), with $t = -b = i$,

$$(26.33) \qquad \phi(q) = (1 + i)(iq)_\infty^{-1} \sum_{n \geq 0} \frac{(-i)^n q^{(n^2+n)/2}}{(q)_n(1 + iq^n)},$$

and from (2.4) and (1.1),

$$(26.34) \qquad \phi(q) = 1 - i \sum_{n \geq 1} \frac{(iq)^n}{(-iq)_n}.$$

In a similar fashion, (12.32), with $\theta = 2\pi/3$, yields

$$(26.41) \qquad\qquad \chi(q) = (1 + \omega^2)F(0, -\omega; -\omega^2),$$

where $\omega = \exp(2\pi i/3)$. From (11.4),

$$(26.42) \qquad \chi(q) = \prod_{n \geq 1} \left(\frac{1 - q^{2n}}{1 + q^{3n}}\right) \left\{1 + \frac{q}{(q)_1} + (1 - q + q^2)\frac{q^2}{(q)_2} + \cdots\right\};$$

from (15.51),

$$(26.43) \qquad \chi(q) = -\omega^2(-\omega^2 q)_\infty^{-1} \sum_{n \geq 0} \frac{\omega^{2n} q^{(n^2+n)/2}}{(q)_n(1 + \omega q^n)};$$

and from (2.4) and (1.1),

$$(26.44) \qquad \chi(q) = 1 - \omega \sum_{n \geq 1} \frac{(-\omega^2 q)^n}{(-\omega q)_n}.$$

The function $\psi(q)$ is somewhat more interesting. It has an obvious interpretation as the generating function for $\beta(n)$, the number of partitions of n into odd parts without gaps, that is, those of the form

$$n = n_1 \cdot 1 + n_3 \cdot 3 + n_5 \cdot 5 + \cdots + n_{2k-1}(2k - 1), \qquad n_j > 0.$$

Now we may write

$$\psi(q^{1/2}) = \sum_{n \geq 1} \frac{(q^{-1/2})^n q^{(n^2+n)/2}}{(q^{1/2})_n} = \sum_{n \geq 1} \frac{q^{n^2/2}}{(q^{1/2})_n}.$$

Comparison with (6.1) yields

$$(26.51) \qquad \psi(q^{1/2}) = (1 - q^{-1/2})F(-1, 0; q^{-1/2}) - 1.$$

By (2.4), therefore,

$$(26.52) \qquad \psi(q^{1/2}) = q^{1/2} F(-1, 0; q^{1/2}),$$

or, replacing q by q^2,

$$(26.53) \qquad \begin{aligned} \psi(q) &= qF(-1, 0; q : q^2) \\ &= q\left\{1 + (1 + q^2)q + (1 + q^2)(1 + q^4)q^2 + \cdots\right\}. \end{aligned}$$

This identity could also have been obtained by reading the conjugate graph, yielding partitions of the form

$$n = p_0 + 2p_1 + 2p_2 + \cdots + 2p_r \qquad (p_0 > p_1 > p_2 > \cdots > p_r),$$

the generating function being the third member of (26.53).

Now, by (20.72), we can obtain

$$(26.54) \qquad 1 + \psi(q) = F(0, q^{-1}; q : q^4),$$

from which can be derived other forms of $\psi(q)$, as above.

We can connect the two arithmetic functions $\beta(n)$ and $P_r(n)$ as follows. Replace q by q^4 in (25.14), then set $t = q$, to get

$$(1 - q)(1 + \psi(q)) = \sum_{r=-\infty}^{+\infty} K_r(q^4)q^r = \sum_{r=-\infty}^{+\infty} \sum_{n} P_r(n)q^{4n+r}.$$

Equating coefficients yields

$$(26.55) \qquad \sum_{4n+r=N} P_r(n) = \beta(N) - \beta(N-1) \qquad (N \geq 2).$$

An alternate form of $\phi(q)$ can be derived from (20.71):

$$(26.56) \qquad \begin{aligned} \phi(q) &= 2F(q^{-1}, 0; -1 : q^2) = 1 + qF(q^{-1}, 0; -q : q^2) \\ &= 1 + q\{1 - (1 - q)q^2 + (1 - q)(1 - q^3)q^4 - \cdots\}. \end{aligned}$$

The coefficients have a rather complicated partition-theoretic interpretation, which we shall not trouble to write down.

To obtain a relation between $\psi(q)$ and $\phi(q)$, we recall (9.1), which we write in the form

$$tF(a, 0; t) = \frac{(aq)_\infty (t^{-1}q)_\infty}{(q)_\infty} \{D(a, t) + R(a, t)\},$$

where

$$D(a, t) = t(1 - aq)^{-1} F(a, aq; t),$$

$$R(a, t) = \sum_{n \geq 1} \frac{(q)_{n-1} q^n}{(aq)_n (t^{-1}q)_n}.$$

Observing that $R(t^{-1}, a^{-1}) = R(a, t)$, we find

$$(26.61) \qquad \begin{aligned} tF(a, 0; t) &- a^{-1}F(t^{-1}, 0; a^{-1}) \\ &= \frac{(aq)_\infty (t^{-1}q)_\infty}{(q)_\infty} \{D(a, t) - D(t^{-1}, a^{-1})\}. \end{aligned}$$

Now we have, by definition, by (6.3), and by (2.2),

$$D(t^{-1}, a^{-1}) = \frac{a^{-1}}{1 - t^{-1}q} F(t^{-1}, t^{-1}q; a^{-1})$$

$$= \frac{a^{-1}}{1 - a^{-1}} F(a^{-1}q^{-1}, a^{-1}; t^{-1}q)$$

$$= \frac{a^{-1}}{1 - a^{-1}} \left\{ 1 + \frac{(1 - a^{-1})t^{-1}q}{1 - a^{-1}q} F(a^{-1}, a^{-1}q; t^{-1}q) \right\}$$

$$= -\frac{1}{1 - a} + a^{-1} D(a^{-1}, t^{-1}q).$$

Thus (26.61) becomes

$$
\begin{aligned}
(26.62) \qquad & tF(a, 0; t) - a^{-1}F(t^{-1}, 0; a^{-1}) \\
& = \frac{(aq)_\infty (t^{-1}q)_\infty}{(q)_\infty} \left\{ \frac{1}{1 - a} + D(a, t) - a^{-1}D(a^{-1}, t^{-1}q) \right\}.
\end{aligned}
$$

Now let $a = -1$, $t = q^{1/2}$, and recall (26.56) and (26.52):

$$\psi(q^{1/2}) + \tfrac{1}{2}\phi(q^{1/2}) = \frac{(-q)_\infty (q^{1/2})_\infty}{(q)_\infty} \left\{ \frac{1}{2} + 2D(-1, q^{1/2}) \right\}.$$

Replacing q by q^2, we have

$$\phi(q) + 2\psi(q) = \frac{(-q^2; q^2)_\infty (q; q^2)_\infty}{(q^2; q^2)_\infty} \left\{ 1 + 4D(-1, q : q^2) \right\}.$$

But

$$
\begin{aligned}
D(-1, q : q^2) &= \sum_{n \geq 1} \frac{q^n}{1 + q^{2n}} = \sum_{n \geq 1} \sum_{m \geq 0} (-1)^m q^{(2m+1)n} \\
&= \sum_{m \geq 0} (-1)^m \frac{q^{2m+1}}{1 - q^{2m+1}} = \sum_{N \geq 1} E_1(N; 4)q^N.
\end{aligned}
$$

Now put $u = \pi/2$ in (9.3). The left side becomes

$$\prod_{n \geq 1} \left(\frac{1 - q^n}{1 + q^n} \right)^2 = \left\{ \sum_{-\infty}^{+\infty} (-1)^n q^{n^2} \right\}^2,$$

and the right side becomes

$$1 + 4 \sum_{N \geq 1} (-q)^N \sum_{\omega | N} \sin \frac{\omega \pi}{2}$$

$$= 1 + 4 \sum_{N \geq 1} (-q)^N \sum_{\omega | N} (-1)^{(\omega - 1)/2}.$$

Replacing q by $-q$ on both sides, we get

$$(26.63) \qquad \left\{ \sum_{-\infty}^{+\infty} q^{n^2} \right\}^2 = 1 + 4 \sum_{N \geq 1} E_1(N; 4)q^N,$$

which is the well-known theorem giving the number of representations of an integer as a sum of two squares.

Thus we get

$$\phi(q) + 2\psi(q) = \prod_{n\geq 1} \frac{(1+q^{2n})(1-q^{2n-1})}{(1-q^{2n})}(1-q^{2n})^2(1+q^{2n-1})^4$$

$$= \prod_{n\geq 1}(1+q^{2n-1})\sum_{-\infty}^{+\infty} q^{m^2}.$$

Replacing q by $-q$ yields

(26.64) $$\phi(-q) + 2\psi(-q) = \frac{1 - 2q + 2q^4 - 2q^9 + \cdots}{(1+q)(1+q^2)(1+q^3)\cdots}.$$

We can also obtain a relation between $\phi(-q)$ and $f(q)$. For this purpose, in (8.1) we set $t = 0$, $u = -q^{1/2}$, $b = -1$, then replace q by q^2 and change q into $-q$. The result is $F + G = HS$, where

$$S = \sum_{n\geq 0}(-1)^n q^{n^2}, \qquad H = (-q)_\infty^{-1},$$

$$G = -\sum_{n\geq 0}\frac{q^{2n+1}}{(-q)_{2n+1}}, \qquad F = \phi(-q).$$

Now

$$G = \tfrac{1}{2}F(0,-1;-q) - \tfrac{1}{2}F(0,-1;q) = 1 - \tfrac{1}{2}f(q) - \tfrac{1}{2}F(0,-1;q),$$

by (26.22). From (2,4) and (6.31), we get

$$F(a,b;q) = (b-aq)^{-1}\left\{\frac{(aq)_\infty}{(bq)_\infty} - (1-b)\right\},$$

so

$$G = \tfrac{1}{2}(-q)_\infty^{-1} - \tfrac{1}{2}f(q).$$

Hence

$$(-q)_\infty^{-1}S = F + G = \phi(-q) + \tfrac{1}{2}(-q)_\infty^{-1} - \tfrac{1}{2}f(q),$$

or

(26.65) $$2\phi(-q) - f(q) = (-q)_\infty^{-1}\sum_{-\infty}^{+\infty}(-1)^n q^{n^2}.$$

Equations (26.64) and (26.65) combine to give the relations stated by Ramanujan in the form

(26.66) $$2\phi(-q) - f(q) = f(q) + 4\psi(-q) = (-q)_\infty^{-1}\sum_{-\infty}^{+\infty}(-1)^n q^{n^2}.$$

Next we establish a relation between $\chi(q)$ and $f(q)$. By (26.41) and (16.4), we may write

(26.71) $$\chi(q) = (q)_\infty^{-1}\left\{1 + \sum_{n\geq 1}\frac{(-1)^n(1+q^n)}{1 - q^n + q^{2n}}q^{(3n^2+n)/2}\right\}.$$

Comparing this with (26.27), we have

$$4\chi(q) - f(q) = 3(q)_\infty^{-1} g(q^3),$$

where

$$g(q) = 1 + 4\sum_{n\geq 1}(-1)^n \frac{q^{(n^2+n)/2}}{1+q^n} = 1 + 4\sum_{N\geq 1} A(N)q^N,$$

the coefficients being given by

$$A(N) = \sum_{\substack{n(n+2k+1)=2N \\ n\geq 1,\ k\geq 0}} (-1)^{n+k}.$$

A slight examination shows that $A(N) = (-1)^N E_1(N;4)$, so

$$g(-q) = \left\{\sum_{-\infty}^{+\infty} q^{n^2}\right\}^2,$$

by (26.63). Therefore

(26.72) $$4\chi(q) - f(q) = 3(q)_\infty^{-1}\left\{\sum_{-\infty}^{+\infty}(-1)^n q^{3n^2}\right\}^2,$$

which is the second equation given by Ramanujan.

In [45], Watson defines three other functions:

(26.81) $\quad \omega(q) = \displaystyle\sum_{n\geq 0} \frac{q^{2n(n+1)}}{\{(1-q)(1-q^3)\cdots(1-q^{2n+1})\}^2},$

(26.82) $\quad \nu(q) = \displaystyle\sum_{n\geq 0} \frac{q^{n(n+1)}}{(1+q)(1+q^3)\cdots(1+q^{2n+1})},$

(26.83) $\quad \rho(q) = \displaystyle\sum_{n\geq 0} \frac{q^{2n(n+1)}}{(1+q+q^2)(1+q^3+q^6)\cdots(1+q^{2n+1}+q^{4n+2})}.$

It is easy to identify these functions. Setting $m = 2$, $r = 1$ in (12.331), we find

(26.84) $$\omega(q) = (1-q)^{-1}F(0, q, q : q^2).$$

We can obtain two forms for $\nu(q)$. First, let $t = -q^{1/2}$, $a = q^{-1/2}$ in (6.1), and then replace q by q^2 to get

(26.85) $$\nu(q) = F(q^{-1}, 0; -q : q^2).$$

By (20.71), therefore,
(26.86)
$$\nu(q) = (1 - iq^{1/2})^{-1}F(0, iq^{1/2}; -iq^{1/2}) = (1 + iq^{1/2})^{-1}F(0, -iq^{1/2}; iq^{1/2}).$$

Similarly, if we set $b = \omega q^{1/2}$, $t = \omega^{-1}q^{1/2}$ in (12.3), where $\omega = \exp(2\pi i/3)$, then replace q by q^2, we have

(26.87) $$\rho(q) = (1 - \omega q)^{-1}F(0, \omega q; \omega^{-1}q : q^2).$$

Now, in (8.4), replace q by q^2 and then let $b = -q$, to get

$$\nu(q) = \prod_{n\geq1} \frac{(1-q^{4n})}{(1-q^{4n-2})^2} - \frac{q}{1-q^2} F(0, q^2; q^2 : q^4)$$

$$= \prod_{n\geq1} \frac{(1-q^{4n})}{(1-q^{4n-2})^2} - q\omega(q^2),$$

which is equivalent to the relation given by Watson:

(26.88) $\qquad \nu(q) + q\omega(q^2) = \nu(-q) - q\omega(q^2) = \prod_{n\geq1}(1+q^{2n})^3(1-q^{2n}).$

It may be of some interest to compare the functions ν, ω with the following:

(26.91) $\qquad \nu^*(q) = \sum_{n\geq0} \frac{(-1)^n q^{n(n+1)}}{(1-q)(1-q^3)\cdots(1-q^{2n+1})},$

(26.92) $\qquad \omega^*(q) = \sum_{n\geq0} \frac{(-1)^n q^n}{(1-q)(1-q^3)\cdots(1-q^{2n+1})}.$

By (12.3), with $b = -t = q^{1/2}$, we find

(26.93) $\qquad \nu^*(q) = (1-q^{1/2})^{-1} F(0, q^{1/2}; -q^{1/2}),$

and by (20.71), with $a = -q^{1/2}$,

(26.94) $\qquad \nu^*(q) = F(q^{-1}, 0; q : q^2).$

But directly from definition, we see that

(26.95) $\qquad \omega^*(q) = (1-q)^{-1} F(0, q; -q : q^2),$

so

(26.96) $\qquad \omega^*(q) = \omega^*(-q) = \nu^*(q^2).$

We can now express $\nu^*(q)$ as a power series of the "false" theta function type. For

$$\nu^*(q) = 1 + q(1-q) F(q, 0; q : q^2),$$

by the transformation $a \to aq$ (4.5). By (7.7),

$$\nu^*(q) = 1 + \sum_{n\geq0}(-1)^n q^{3n^2+4n+1} + \sum_{n\geq1}(-1)^n q^{3n^2+2n}.$$

Hence

(26.97) $\qquad \nu^*(q) = 1 + \sum_{n\geq1}(-1)^n q^{3n^2}(q^{2n} - q^{-2n}).$

27. The functions $L(N)$ and $J(N)$. In the preceding section we defined $J(N)$ by

(27.1) $\qquad \sum_{N\geq1} J(N)q^N = \sum_{n\geq1}(-1)^n \frac{q^{(3n^2+n)/2}}{1+q^n},$

and showed that for $N \geq 1$,

$$(27.11) \qquad 2J(N) = L(N) + \varepsilon(N) - 2 \sum_{1 \leq m < \sqrt{N}} (-1)^{m-1} L(N - m^2).$$

We shall now examine $J(N)$ a little more closely. By expanding the factor $(1 + q^n)^{-1}$ in (27.1), we have

$$(27.12) \qquad \sum_{N \geq 1} J(N) q^N = \sum_{N \geq 1} q^N \sum_{\substack{3n^2 + (2k+1)n = 2N \\ n \geq 1, \ k \geq 0}} (-1)^{n+k},$$

so that

$$(27.13) \qquad J(N) = \sum_{\substack{n \mid 2N \\ n(3n+\omega) = 2N}} (-1)^{n + (\omega - 1)/2},$$

where ω denotes a positive odd integer. Now let $N = 2^\alpha m$, m odd. Then n must be of the form $2^\beta \mu$, with $\mu \mid m$ and $\beta \leq \alpha + 1$. It is clear that $\beta = 0$, $\alpha + 1$ are the only possibilities. Also, $2N \geq n(3n+1)$, so $(6n+1)^2 \leq 24N + 1$. Conversely, every divisor of $2N$ which satisfies these two conditions is admissible. Hence
(27.14)
$$J(N) = \sum_{\substack{\mu \mid m \\ (6\mu+1)^2 \leq 24N+1}} (-1)^{\mu + (\omega-1)/2} + \sum_{\substack{\mu \mid m \\ (12 \cdot 2^\alpha \mu + 1)^2 \leq 24N+1}} (-1)^{2^{\alpha+1} \mu + (\omega-1)/2}$$

$$= - \sum_{\substack{\mu \mid m \\ (6\mu+1)^2 \leq 24N+1}} (-1)^{(\omega-1)/2} + \sum_{\substack{\mu \mid m \\ (12 \cdot 2^\alpha \mu + 1)^2 \leq 24N+1}} (-1)^{(\omega-1)/2}.$$

In the first sum,

$$\omega = 2N/n - 3n \equiv 2^{\alpha+1}(m/\mu) + \mu \pmod 4.$$

Therefore $(\omega - 1)/2 \equiv N + (\mu - 1)/2 \pmod 2$, and the first sum is

$$(-1)^{N+1} \sum_{\substack{\mu \mid m \\ (6\mu+1)^2 \leq 24N+1}} (-1)^{(\mu-1)/2}.$$

In the second sum,

$$\omega = 2N/n - 3n \equiv m/\mu + 2^{\alpha+1}\mu$$
$$\equiv m\mu + 2N \equiv m + \mu - 1 + 2N \pmod 4.$$

Hence the second sum is

$$(-1)^{N+(m-1)/2} \sum_{\substack{\mu \mid m \\ (12 \cdot 2^\alpha \mu + 1)^2 \leq 24N+1}} (-1)^{(\mu-1)/2}.$$

Thus we have, for $N = 2^\alpha m$, m odd,

$$
\text{(27.15)} \quad J(N) = (-1)^{N+1} \left\{ \sum_{\substack{\mu \mid m \\ (6\mu+1)^2 \le 24N+1}} (-1)^{(\mu-1)/2} \right.
$$

$$
\left. -(-1)^{(m-1)/2} \sum_{\substack{\mu \mid m \\ (12 \cdot 2^\alpha \mu+1)^2 \le 24N+1}} (-1)^{(\mu-1)/2} \right\}.
$$

It is now easy to consider some special cases. For example, let $N = p$, a prime ≥ 7. In both sums $\mu = 1$ is admissible and $\mu = p$ is not. Therefore

$$
\text{(27.16)} \qquad J(p) = 1 - (-1)^{(p-1)/2}.
$$

In order to use (27.11), we observe that the sum

$$
\text{(27.17)} \qquad s(N) = \sum_{1 \le m < \sqrt{N}} (-1)^{m-1} L(N - m^2)
$$

is usually even. For we get an odd summand $L(n - m^2)$ only if $N = m^2 + k^2$, in which case $L(N - k^2)$ is also odd, so that the representations of N as a sum of two squares pair off, unless $m = k$, in which case $s(N)$ is odd. Hence

$$
\text{(27.18)} \qquad s(N) \equiv \begin{cases} 1 \pmod 2 & \text{if } N = 2m^2, \\ 0 \pmod 2 & \text{if } N \ne 2m^2. \end{cases}
$$

Combining the results of (27.11), (27.16), and (27.18), we find that

$$
\text{(27.19)} \qquad L(p) \equiv 0 \pmod 4 \qquad (p \text{ prime}, \ge 7).
$$

Now let $N = 2^\alpha$, $\alpha > 0$. The only possible divisor, $\mu = 1$, is admissible in the first sum but not in the second. Therefore

$$
\text{(27.2)} \qquad J(2^\alpha) = -1 \qquad (\alpha > 0).
$$

If α is even, $\varepsilon(2^\alpha) = 1$, $s(2^\alpha) \equiv 0 \pmod 2$, so

$$
\text{(27.21)} \qquad L(2^{2k}) \equiv 1 \pmod 4 \qquad (k \ge 0).
$$

If α is odd, $\varepsilon(2^\alpha) = 0$, $s(2^\alpha) \equiv 1 \pmod 2$, so

$$
\text{(27.22)} \qquad L(2^{2k+1}) \equiv 0 \pmod 4 \qquad (k \ge 0).
$$

Now suppose that $N = p^{2\gamma}$, where p is an odd prime, and $\gamma > 0$. It is easily verified that, in (27.15), μ runs over the values $1, p, p^2, \ldots, p^{\gamma-1}$ exactly in both sums. Since $p^{2\gamma} \equiv 1 \pmod 4$, we have

$$
\text{(27.23)} \qquad J(p^{2\gamma}) = 0 \qquad (p \text{ an odd prime}, \gamma > 0).
$$

Next consider $N = p^{2\gamma-1}$, $\gamma > 0$. For $p \ge 7$, we find again that μ runs over $1, p, p^2, \ldots, p^{\gamma-1}$, so

$$
\text{(27.24)} \qquad J(p^{2\gamma-1}) = (1 - (-1)^{(p-1)/2}) \sum_{k=0}^{\gamma-1} (-1)^{(p^k-1)/2}.
$$

Hence, if $p \equiv 1 \pmod 4$, $J(p^{2\gamma-1}) = 0$; if $p \equiv -1 \pmod 4$,

$$J(p^{2\gamma-1}) = 2 \sum_{k=0}^{\gamma-1} (-1)^k = 1 - (-1)^\gamma,$$

and

$$J(p^{4\delta-1}) = 0, \qquad J(p^{4\delta-3}) = 2.$$

It is not difficult to show that

$$J(3^{2\gamma-1}) = J(5^{2\gamma-1}) = 1.$$

We may summarize our results as follows:

(27.25)   If p is an odd prime and $\alpha > 0$, then $J(p^\alpha) = 0$ except when
          (i) $p \equiv -1 \pmod 4$, $\alpha \equiv 1 \pmod 4$, $p \geq 7$; then $J(p^\alpha) = 2$,
          (ii) $p = 3, 5$, α odd; then $J(p^\alpha) = 1$.

It follows directly from (27.25), (27.11), and (27.17) that

(27.26)   If p is an odd prime, $\alpha > 0$, then (i) $L(p^{2\alpha}) \equiv 1 \pmod 4$, (ii)
          $L(p^{2\alpha-1}) \equiv 0 \pmod 4$ if $p \geq 7$, (iii) $L(3^{2\alpha-1}) \equiv L(5^{2\alpha-1}) \equiv 2$
          $\pmod 4$.

By similar arguments, we can show that if r and s are odd primes satisfying (a) $s > 6r$, (b) $r \equiv s \pmod 4$, then

(27.27) $$L(rs) = 2 \sum_{m \geq 1} (-1)^{m-1} L(rs - m^2) \equiv 0 \pmod 4.$$

Some of the results of this section were stated in [16].

Notes

The mock theta functions have made several appearances in the literature in recent years [1, 2, 3, 4, 5, 6, 7]. The results in this chapter all relate to the third-order mock theta functions whose basic properties were laid out by G. N. Watson [8]. The extensive treatment of the mock theta functions by Ramanujan in his "Lost" Notebook [5] suggests that these functions possess a rich mathematical structure and are still far from understood.

REFERENCES

1. R. P. Agarwal, *Certain basic hypergeometric identities associated with mock theta functions*, Quart. J. Math. Oxford Ser. (2) **20** (1969), 121–128.
2. G. E. Andrews, *On basic hypergeometric functions, mock theta functions and partitions.* (I), Quart. J. Math. Oxford Ser. (2) **17** (1966), 64–80.
3. _____, *On basic hypergeometric functions, mock theta functions and partitions* (II), Quart. J. Math. Oxford Ser. (2) **17** (1966), 132–143.
4. _____, *On the theorems of Watson and Dragonette for Ramanujan's mock theta functions*, Amer. J. Math. **88** (1966), 454–490.
5. _____, *An introduction to Ramanujan's "lost" notebook*, Amer. Math. Monthly **86** (1979), 89–108.

6. _____ , *Combinatorics and Ramanujan's "lost" notebook*, Surveys in Combinatorics, I. Anderson, editor, London Math. Soc. Lecture Notes Series, no. 103, Cambridge Univ. Press, 1985, pp. 1–23.

7. _____ , *The fifth and seventh order mock theta functions*, Trans. Amer. Math. Soc. **293** (1986), 113–134.

8. G. N. Watson, *The final problem: An account of the mock theta functions*, J. London Math. Soc. **11** (1936), 55–80.

CHAPTER 4

Other Applications

28. Theorems of Liouville type. In a series of articles [29] Liouville stated many identities for general functions satisfying certain parity conditions. When specialized these yield results of number-theoretic interest. It is known that the Liouville identities are equivalent to identities among elliptic functions, and methods of paraphrasing the latter to obtain the former have been used by various authors. We shall apply this method first to (9.3).

By using Jacobi's product, we see easily that (9.3) may be written in the form

$$(28.1) \quad (q)_\infty^3 \left\{ \sum_{n\geq 0} (-1)^n \frac{\sin(2n+1)u}{\sin u} q^{(n^2+n)/2} \right\}^{-1}$$
$$= 1 - 4 \sin u \sum_{N\geq 1} q^N \sum_{\omega \mid N} \sin\left(\frac{2N}{\omega} - \omega\right) u.$$

If we let $u \to 0$, we find the well-known formula of Jacobi:

$$(28.11) \quad (q)_\infty^3 = \sum_{n\geq 0} (-1)^n (2n+1) q^{(n^2+n)/2}.$$

Hence (28.1) may be written as

(28.12)
$$\sum_{n\geq 0} (-1)^n q^{(n^2+n)/2} \left\{ \sin(2n+1)u - (2n+1)\sin u \right\}$$

$$= 4 \sin u \sum_{n\geq 0} (-1)^n \sin(2n+1)u \; q^{(n^2+n)/2} \cdot \sum_{N\geq 1} q^N \sum_{\omega \mid N} \sin\left(\frac{2N}{\omega} - \omega\right) u$$

$$= \sum_{M\geq 1} q^M C_M(u),$$

where

$$(28.13) \quad C_M(u) = 4 \sum (-1)^n \sin u \sin(2n+1)u \sin\left(\frac{2N}{\omega} - \omega\right) u,$$

the summation being extended over all values of n, ω, N satisfying

$$(28.14) \quad n \geq 0, \quad N \geq 1, \quad \omega \text{ odd} > 0, \quad \omega \mid N, \quad \frac{n^2+n}{2} + N = M.$$

Let us define $\omega_1 = 2n + 1$, $\omega_2 = \omega$, $d = N/\omega_2$, $k = 8M + 1$. Then (28.14) becomes

(28.15) $\qquad\qquad \omega_1, \omega_2$ odd > 0, $\quad d > 0$, $\quad k = \omega_1^2 + 8d\omega_2$,

and (28.13) becomes

(28.16) $\quad D_k(u) \equiv C_{(k-1)/8}(u) = 4 \sum (-1)^{(\omega_1-1)/2} \sin u \sin \omega_1 u \sin(2d - \omega_2)u.$

By elementary trigonometry we can write the summand as a sum of sines:
(28.17)
$$D_k(u) = \sum (-1)^{(\omega_1-1)/2}\{\sin(\omega_1 + \omega_2 - 2d + 1)u + \sin(\omega_1 - \omega_2 + 2d - 1)u$$
$$- \sin(\omega_1 + \omega_2 - 2d - 1)u - \sin(\omega_1 - \omega_2 + 2d + 1)u\}.$$

Replacing q by q^8 in (28.12) and multiplying by q, we get

(28.18) $\displaystyle\sum_{n\geq 0} (-1)^n q^{(2n+1)^2}\{\sin(2n+1)u - (2n+1)\sin u\} = \sum_{\substack{k>0 \\ k\equiv 1 \,(\mathrm{mod}\, 8)}} q^k D_k(u).$

Equating coefficients of q^k, we find that $D_k(u) = 0$ unless $k = r^2$, and if $k = r^2$,

(28.19) $\qquad\qquad D_{r^2}(u) = (-1)^{(r-1)/2}\{\sin ru - r\sin u\}.$

Thus we have obtained a set of identities in u of the form

(28.2) $\qquad\qquad \displaystyle\sum_j A_j^{(k)} \sin ju = 0 \qquad (k = 1, 9, 17, \dots).$

It follows that $A_j^{(k)} - A_{-j}^{(k)} = 0$ for all j and k, so we may replace $\sin ju$, whenever it appears, by an arbitrary odd function of the integral variable j. We have therefore proved the following theorem.

(28.3) *Let k be any positive integer of the form $8n + 1$, let $f(j)$ be any odd function of j, and let S_k denote the sum*

$$\sum (-1)^{(\omega_1-1)/2}\{f(\omega_1 + \omega_2 - 2d + 1) + f(\omega_1 - \omega_2 + 2d - 1)$$
$$- f(\omega_1 + \omega_2 - 2d - 1) - f(\omega_1 - \omega_2 + 2d + 1)\},$$

extended over all solutions (ω_1, ω_2, d) of

$$k = \omega_1^2 + 8d\omega_2,$$

where ω_1, ω_2 are odd and positive, d is positive. Then

$$S_k = \begin{cases} 0 & \text{if } k \text{ is not a square,} \\ (-1)^{(r-1)/2}(f(r) - rf(1)) & \text{if } k = r^2. \end{cases}$$

Another way of paraphrasing (28.12) is to divide the first two members by $\sin u$, express $\sin(2n + 1)u \sin(2N/\omega - \omega)u$ on the left as a difference of two cosines, and use the identity

$$\frac{\sin(2n + 1)u}{\sin u} = 1 + 2\cos 2u + \cdots + 2\cos 2nu$$

on the right. We shall not carry through the details, but merely state the result.

(28.4) *Let k be any positive integer of the form $8n+1$, let $g(j)$ be any even function of j, and let T_k denote the sum*

$$\sum (-1)^{(\omega_1-1)/2}\{g(\omega_1-\omega_2+2d)-g(\omega_1+\omega_2-2d)\},$$

extended over all solutions (ω_1,ω_2,d) of

$$k = \omega_1^2 + 8d\omega_2,$$

where ω_1,ω_2 are odd and positive, d is positive. Then

$$T_k = \begin{cases} 0 & \text{if } k \text{ is not a square.} \\ (-1)^{(r-1)/2}\left\{\dfrac{r-1}{2}g(0) - \sum_{j=1}^{(r-1)/2} g(2j)\right\} & \text{if } k = r^2,\ r > 0. \end{cases}$$

29. Another identity and paraphrase. In (26.61), put $t = a$ and use (7.7); the left side becomes

(29.1)

$$aF(a,0;a) - a^{-1}F(a^{-1},0;a^{-1})$$

$$= \frac{a}{1-a}\left\{\sum_{n\geq 0}(-1)^n q^{(3n^2+n)/2}a^{3n} + \sum_{n\geq 1}(-1)^n q^{(3n^2-n)/2}a^{3n-1}\right\}$$

$$- \frac{a^{-1}}{1-a^{-1}}\left\{\sum_{n\geq 0}(-1)^n q^{(3n^2+n)/2}a^{-3n} + \sum_{n\geq 1}(-1)^n q^{(3n^2-n)/2}a^{-3n+1}\right\}.$$

With $a = \exp(2iu)$, this becomes

(29.11)
$$\frac{2e^{iu}}{1-e^{2iu}} \sum_{n=-\infty}^{+\infty} (-1)^n \cos(6n+1)u\, q^{(3n^2+n)/2}.$$

On the right side of (26.61), an application of Jacobi's identity yields

(29.12)
$$\frac{(aq)_\infty (a^{-1}q)_\infty}{(q)_\infty} = -\frac{2ie^{iu}}{1-e^{2iu}}(q)_\infty^{-2} \sum_{n\geq 0}(-1)^n \sin(2n+1)u\, q^{(n^2+n)/2}.$$

Now it is easy to see that

(29.13)
$$D(a,a) = \frac{a}{1-a} + \sum_{n\geq 1}q^n \sum_{d\mid n} a^{d+n/d},$$

so that, with $a = \exp(2iu)$,

(29.14)
$$D(a,a) - D(a^{-1},a^{-1}) = i\cot u + 2i\sum_{n\geq 1}q^n \sum_{d\mid n} \sin 2\left(d+\frac{n}{d}\right)u.$$

Combining (29.11), (29.12), and (29.14), we obtain

$$(q)_\infty^2 \sum_{m=-\infty}^{+\infty} (-1)^m \cos(6m+1)u \; q^{(3m^2+m)/2}$$

(29.15)
$$= \left\{ \sum_{n \geq 0} (-1)^n \sin(2n+1)u \; q^{(n^2+n)/2} \right\}$$

$$\times \left\{ \cot u + 2 \sum_{m \geq 1} q^m \sum_{d \mid m} \sin 2 \left(d + \frac{m}{d} \right) u \right\}.$$

If we compare (29.15) with (28.1), then paraphrase the resulting identity, we obtain the following theorem:

(29.2) *Let $F(r)$ be any odd function, $M \equiv 1 \pmod{24}$; let d, d' be positive, ω odd and positive, $u \equiv 1 \pmod 6$, positive or negative.*

Define

$$G(M) = \sum_{(a)} (-1)^{(u-1)/6} F(d+d')$$

$$+ \sum_{(b)} (-1)^{(u-1)/6} \left\{ F\left(d + \frac{u-\omega}{2} \right) + F\left(d - \frac{u+\omega}{2} \right) \right\},$$

where the sums are taken over all solutions of

$$\text{(a)} \quad M = u^2 + 24dd',$$
$$\text{(b)} \quad M = u^2 + 24d\omega.$$

Then

$$G(M) = \begin{cases} 0 & \text{unless } M \text{ is a square,} \\ -(-1)^{(s-1)/6} \displaystyle\sum_{1 \leq r \leq (|s|-1)/2} F(r) & \text{if } M = s^2, \; s \equiv 1 \pmod 6. \end{cases}$$

An alternate paraphrasing of (29.15) leads to:

(29.3) *Let $F(r)$ be any even function, $M = 24N + 4$, $N > 0$; let ω, ω' be odd and positive, $\varepsilon, \varepsilon'$ even and positive. Then*

$$\sum_{(a)} (-1)^{(\omega-1)/2+(\omega'\pm1)/6} \{ F(\omega+\varepsilon+\varepsilon') - F(\omega-\varepsilon-\varepsilon') \}$$

$$= \sum_{(b)} (-1)^{(\omega-1)/2+(\omega'\pm1)/6} \left\{ 2 \sum_{\substack{0 < u \leq \omega \\ u \text{ odd}}} F(u) - \omega F(\omega') - F(\omega) \right\},$$

where $\pm$ in the exponents are chosen to make $(\omega' \pm 1)/6$ an integer, and the sums are taken over all solutions of

$$\text{(a)} \quad M = 3\omega^2 + \omega'^2 + 6\varepsilon\varepsilon',$$
$$\text{(b)} \quad M = 3\omega^2 + \omega'^2.$$

30. Two more identities and their paraphrases. Starting with the function

$$(30.1) \qquad h(a) = (1+a) \prod_{n\geq 1} \frac{(1-a^2 q^n)(1-a^{-2}q^n)}{(1-aq^n)(1-a^{-1}q^n)},$$

it is easy to show that $h(a) = ah(a^{-1})$ and that $h(a) = -a^3 qh(aq)$. From these, together with $h(1) = 2$, we obtain

$$(30.11) \qquad (q)_\infty h(a) = a^{1/2} \sum_{n=-\infty}^{+\infty} (-1)^n q^{(3n^2+n)/2} \{a^{3n+1/2} + a^{-3n-1/2}\}.$$

By Jacobi's product, however, we have

$$(30.12) \qquad h(a) = \frac{\sum_{-\infty}^{+\infty}(-a^2)^n q^{(n^2-n)/2}}{\sum_{-\infty}^{+\infty}(-a)^n q^{(n^2-n)/2}}.$$

Writing $a = \exp(2iu)$ in (30.11) and (30.12), then equating both expressions for $h(a)$, we obtain easily

(30.2)
$$\sum_{-\infty}^{+\infty}(-1)^m q^{(3m^2+m)/2} \sum_{n\geq 0}(-1)^n q^{(n^2+n)/2} \sin(4n+2)u$$

$$= 2\sum_{-\infty}^{+\infty}(-1)^m q^{(3m^2+m)/2}\cos(6m+1)u \sum_{n\geq 0}(-1)^n q^{(n^2+n)/2}\sin(2n+1)u.$$

Paraphrasing this identity yields the following theorem:

(30.3) *Let $F(r)$ be any odd function; $N \equiv 4 \pmod{24}$; ω odd and positive; $t \equiv 1 \pmod 6$, t positive or negative. Then*

$$\sum_{(a)}(-1)^{(t+\omega)/2}\left\{F\left(\frac{\omega+t}{2}\right) + F\left(\frac{\omega-t}{2}\right) - F(\omega)\right\} = 0,$$

where the sum is taken over all solutions of

$$(a) \quad N = t^2 + 3\omega^2.$$

As a final example in this category, we give a paraphrase of (18.7):

(30.4) *Let $F(r)$ be any odd function, and let d_1, d_2, d_3, d_4, d_5 denote positive integers, $\omega_1, \omega_2, \omega_3$, odd and positive integers, satisfying*

$$(a) \quad N = d_1\omega_1 + d_2\omega_2,$$
$$(30.41) \qquad (b) \quad N = d_3\omega_3,$$
$$(c) \quad N = d_4 d_5.$$

Then

$$\sum_{(a)}\{2F(2d_1 - \omega_1 - 2d_2 + \omega_2 + 1) - F(2d_1 - \omega_1 + 2d_2 - \omega_2 + 1)$$

$$- F(-2d_1 + \omega_1 - 2d_2 + \omega_2 + 1)\}$$

(30.42)
$$= 2\sum_{(b)} F(2d_3 - \omega_3) - \sum_{(c)} d_4\{F(2d_4 - 2d_5 + 1) + F(-2d_4 + 2d_5 + 1)\}.$$

Various specializations of the theorems proved in §§28–30 lead to interesting results in number theory, but we shall not take the space to state them here.

31. Sums of squares. The reader will already have observed that many of these identities that we have proved lead, upon suitable specialization, to theorems about the number of representations of integers by certain quadratic forms. In order to avoid lengthy digressions, we have refrained from stating those theorems in their context, and have collected them here, with brief indications of their proofs. Most of the results in this section are known, e.g. [**33**], but we believe that some of the derivations given here are as direct as any in the literature. See Grosswald [**19**].

In (26.63), we have proved

$$\text{(31.1)} \qquad \sum 1 \; (N = a^2 + b^2) = 4E_1(N; 4).$$

For our next example, put $k = 2$, $\alpha = -1$ in (19.5):

$$\text{(31.11)} \qquad \prod_{n \geq 1} \frac{(1 - q^{pn})^2(1 + q^{pn-r})(1 + q^{pn-p+r})}{(1 + q^{pn})^2(1 - q^{pn-r})(1 - q^{pn-p+r})}$$

$$= 1 + 2 \sum_{N \geq 1} E_{r,p-r}(N; 2p)q^N.$$

With $p = 2$, $r = 1$, we recover (31.1). With $p = 4$, $r = 1$, the product may be transformed into

$$\prod_{n \geq 1}(1 - q^{4n})(1 + q^{4n-2})^2(1 - q^{2n})(1 + q^{2n-1})^2 = \sum_{-\infty}^{+\infty} q^{2a^2+b^2},$$

by Jacobi's identity. Hence

$$\text{(31.12)} \qquad \sum 1 \; (N = 2a^2 + b^2) = 2E_{1,3}(N; 8).$$

In (18.6), put $t = -1$. Then

$$\text{(31.13)} \qquad S(b, q) \equiv \frac{1 + b}{1 - b} + \sum_{k,n \geq 1}(-1)^n q^{kn}(b^k - b^{-k})$$

$$= \prod_{n \geq 1} \frac{(1 - q^n)^2(1 + bq^{n-1})(1 + b^{-1}q^n)}{(1 + q^n)^2(1 - bq^{n-1})(1 - b^{-1}q^n)}.$$

As a function of b, the even part of $S(b, q)$ is

$$\text{(31.14)} \qquad \frac{1}{2}(S(b, q) + S(-b, q)) = \frac{1 + b^2}{1 - b^2} + 2 \sum_{k,n \geq 1}(-1)^n q^{2kn}(b^{2k} - b^{-2k})$$

$$= S(b^2, q^2).$$

Now $S(q^r, q^p)$ is given by (31.11). Hence

$$S(-q^r, q^p) = 2S(q^{2r}, q^{2p}) - S(q^r, q^p),$$

which may be written in the form

$$\prod_{n\geq 1} \frac{(1-q^{pn})^2(1-q^{pn-r})(1-q^{pn-p+r})}{(1+q^{pn})^2(1+q^{pn-r})(1+q^{pn-p+4})}$$

(31.15)

$$= 1 - 2\sum_{N\geq 1} q^N\{E_{r,p-r}(N;2p) - 2E_{r,p-r}(N/2;2p)\}.$$

Putting $p = 3$, $r = 1$, we have

$$\prod_{n\geq 1} \frac{(1-q^{3n})(1-q^n)}{(1+q^{3n})(1+q^n)} = \sum_{-\infty}^{+\infty}\sum_{-\infty}^{+\infty}(-1)^{a+b}q^{3a^2+b^2}$$

(31.16)

$$= 1 - 2\sum_{N\geq 1} q^n\{E_{1,2}(N;6) - 2E_{1,2}(N/2/;6)\}.$$

Replacing q by $-q$, we obtain the result

(31.17)
$$\sum 1\ (N = 3a^2 + b^2) = 2u(N),$$

where

(31.18)
$$u(N) = \begin{cases} E_1(N;3), & N \text{ odd,} \\ 3E_1(N;3), & N \text{ even.} \end{cases}$$

Next, in (19.4), with $\alpha = 1$, we have

$$\prod_{n\geq 1} \frac{(1-q^{pn})^2(1-q^{pn-r-s})(1-q^{pn-p+r+s})}{(1-q^{pn-r})(1-q^{pn-p+r})(1-q^{pn-s})(1-q^{pn-p+s})}$$

(31.19)

$$= \sum_{N\geq 0} E_r(pN + rs;p)q^N.$$

For $p = 4$, $r = s = 1$, this becomes

(31.2)
$$\prod_{n\geq 1} \frac{(1-q^{2n})^2}{(1-q^{2n-1})^2} = \sum_{N\geq 0} E_1(4N + 1;4)q^N,$$

which by (7.321) is equivalent to

$$\sum 1\left(N = \frac{a^2+a}{2} + \frac{b^2+b}{2}; a,b \geq 0\right) = E_1(4N + 1;4),$$

(31.21) $\quad \sum 1\ (8N + 2 = \omega_1^2 + \omega_2^2; \omega_1, \omega_2 \text{ odd} > 0) = E_1(4N + 1;4).$

Similarly, for $p = 6$, $r = 1$, $s = 3$, we find

(31.22) $\quad \sum 1\ (8N + 4 = \omega_1^2 + 3\omega_2^2; \omega_1, \omega_2 \text{ odd} > 0) = E_1(2N + 1;3).$

For $p = 8$, $r = 1$, $s = 3$, we have

(31.23) $\quad \sum 1\ (8N + 3 = \omega_1^2 + 2\omega_2^2; \omega_1, \omega_2 \text{ odd} > 0) = E_1(8N + 3;8),$

and for $p = 8$, $r = 1$, $s = 5$,

(31.24) $\quad \sum 1\ (8N + 5 = \omega_1^2 + 4\omega_2^2; \omega_1, \omega_2 \text{ odd} > 0) = E_1(8N + 5;8).$

Undoubtedly some of these results can be obtained directly from (31.1), (31.12), and (31.17).

Next we consider some examples of forms in four variables. In (18.62), letting $u = \pi$, we obtain

$$(31.3) \qquad \prod_{n \geq 1} \left(\frac{1 - q^n}{1 + q^n} \right)^4 = 1 - 8 \sum_{N \geq 1} q^N \sum_{nk=N} (-1)^{k+n} n,$$

so that, after q is replaced by $-q$,

$$(31.31) \qquad \sum 1 \, (N = a^2 + b^2 + c^2 + d^2) = 8(2 + (-1)^N) \sum_{\substack{w \mid N \\ w \text{ odd}}} w.$$

Similarly, putting $u = \pi/2$ yields

$$(31.32) \qquad \sum 1 \, (2^\alpha m = a^2 + b^2 + 2c^2 + 2d^2) = 4k(\alpha)\sigma(m),$$

where m is odd, $\sigma(m)$ is the sum of the divisors of m, and $k(0) = 1$, $k(1) = 2$, $k(\alpha) = 6$ for $\alpha \geq 2$.

To deal with the next case, $N = a^2 + b^2 + c^2 + 2d^2$, we start with (9.3), differentiate logarithmically with respect to u, and rearrange to get

$$(31.33)$$
$$1 + 4 \sin u \tan u \sum_{n \geq 1} C_N(u) q^N$$
$$= \prod_{n \geq 1} \frac{(1 - q^n)^2}{(1 - 2q^n \cos 2u + q^{2n})} \left\{ 1 + 4 \tan u \sum_{N \geq 1} q^N \sum_{d \mid N} \sin 2du \right\},$$

where

$$C_N(u) = \sum_{\substack{w \mid N \\ w \text{ odd}}} \left(\frac{2N}{w} - w \right) \cos \left(\frac{2N}{w} - w \right) u.$$

Now put $u = \pi/4$. The right side of (31.33) becomes

$$\prod_{n \geq 1} \frac{(1 - q^n)^2}{(1 + q^{2n})} \left\{ 1 + 4 \sum_{N \geq 1} E_1(N; 4) q^N \right\} = \prod_{n \geq 1} \frac{(1 - q^n)^2}{(1 + q^{2n})} \left\{ \sum_{-\infty}^{+\infty} q^{n^2} \right\}^2$$

$$= \prod_{n \geq 1} \frac{(1 - q^n)^2 (1 - q^{2n})^2 (1 + q^{2n-1})^4}{(1 + q^{2n})}$$

$$= \prod_{n \geq 1} \{(1 - q^{4n})(1 - q^{4n-2})^2\}^3 (1 - q^{2n})(1 + q^{2n-1})^2$$

$$= \sum_{-\infty < m,n,r,s < +\infty} (-1)^{n+r+s} q^{m^2 + 2n^2 + 2s^2 + 2r^2}.$$

Denoting this function by $g(q)$, we see that the even part of $g(q)$ is

$$\frac{1}{2}(g(q) + g(-q)) = \sum_{m,n,r,s} (-1)^{n+r+s} q^{4m^2 + 2n^2 + 2r^2 + 2s^2} = h(q^2),$$

say. Now

$$h(-q) = \sum_{N \geq 0} A(N)q^N,$$

where

$$A(N) = \sum 1 \ (N = a^2 + b^2 + c^2 + 2d^2).$$

Thus, if we write $D_N = 2\sqrt{2}C_N(\pi/4)$ for $N \geq 1$, $D_0 = 1$, we have

$$\sum_{N \geq 0} A(N)q^{2N} = h(-q^2) = h((iq)^2) = \frac{1}{2}(g(iq) + g(-iq))$$

$$= \sum_{N \geq 0} D_N \left(\frac{i^N + i^{-N}}{2} \right) q^N = \sum_{N \geq 0} (-1)^N D_{2N} q^{2N}.$$

Hence $A(N) = (-1)^N D_{2N}$. Now for $n > 0$,

$$\frac{1}{2}D_n = \sqrt{2}C_n = \sum_{\omega \mid n} \left(\frac{2n}{\omega} - \omega \right) \psi \left(\frac{2n}{\omega} - \omega \right),$$

where $\psi(2k) = 0$, $\psi(8k \pm 1) = 1$, $\psi(8k \pm 3) = -1$. We observe that ψ is totally multiplicative, that is, $\psi(kl) = \psi(k)\psi(l)$ for all positive integers k, l. Further, $2n/\omega \equiv 2n\omega \pmod{8}$, so

$$\frac{1}{2}D_n = \sum_{d \mid n} \left(\frac{2n}{d} - d \right) \psi(2nd - d)$$

$$= \psi(2n - 1) \sum_{d \mid n} \left(\frac{2n}{d} - d \right) \psi(d).$$

Now if $n = 2^\alpha m$, m odd,

$$\frac{1}{2}D_n = \psi(2n - 1) \left\{ 2^{\alpha+1} \sum_{d \mid m} \frac{m}{d} \psi(d) - \sum_{d \mid m} d\psi(d) \right\}$$

$$= \psi(2n - 1) \left\{ 2^{\alpha+1} \sum_{d \mid m} d\psi \left(\frac{m}{d} \right) - \sum_{d \mid m} d\psi(d) \right\}$$

$$= \psi(2n - 1) \left\{ 2^{\alpha+1}\psi(m) - 1 \right\} \sum_{d \mid n} d\psi(d).$$

The sum is a multiplicative function of n; that is, it satisfies $f(mn) = f(m)f(n)$ if $(m, n) = 1$. It is easy to see that for $n = 2^\beta$ it has the value 1, and for $n = p^\beta$, p an odd prime, it has the value

$$\frac{(p\psi(p))^{\beta+1} - 1}{p\psi(p) - 1} \neq 0.$$

Thus D_n is never 0, so $A(N)$ is always positive. We have therefore proved that $a^2 + b^2 + c^2 + 2d^2$ is a universal form. Explicitly, if $N = 2^\alpha m$, m odd,

$$A(N) = 2(-1)^N \psi(4N - 1)\{2^{\alpha+2}\psi(m) - 1\} \sum_{d \mid N} d\psi(d).$$

But $\psi(4N - 1) = (-1)^N$, so

(31.34) $$A(N) = 2\{2^{\alpha+2}\psi(m) - 1\} \sum_{d \mid N} d\psi(d).$$

In (18.85), put $p = 6$, $s = 1$, $a = 3$. Then the product reduces to

(31.4) $$\prod_{n \geq 1} \frac{(1 - q^n)^2(1 - q^{3n})^2}{(1 + q^n)^2(1 + q^{3n})^2} = \sum (-1)^{a+b+c+d} q^{a^2+b^2+3c^2+3d^2},$$

and the sum to

(31.41) $$1 - 4 \sum_{n \geq 1} \frac{n\chi(n)q^n}{1 - q^n},$$

where

(31.42) $$\chi(n) = \begin{cases} +1, & n \equiv \pm 1 \ (\mathrm{mod}\,6), \\ -1, & n \equiv \pm 2 \ (\mathrm{mod}\,6), \\ 0, & n \equiv 0 \ (\mathrm{mod}\,3). \end{cases}$$

If we replace q by $-q$ and equate coefficients, there results

(31.43) $$\sum 1 \ (N = a^2 + b^2 + 3c^2 + 3d^2) = (-1)^{N-1} 4 \sum_{d \mid N} d\chi(d).$$

In (18.62), replace q by q^p, e^{iu} by q^r, with $0 < r < p$, $(r, p) = 1$. After the usual simplifications, we find

(31.5) $$q^r \left\{ \prod_{n \geq 1} \frac{(1 - q^{pn})^2}{(1 - q^{pn-r})(1 - q^{pn-p+r})} \right\}^2 = \frac{1}{p} \sum_{N \geq 1} q^N \sum_{\substack{d\delta = pN - r^2 \\ d \equiv r \ (\mathrm{mod}\,p)}} (d + \delta).$$

For $p = 2$, $r = 1$, this yields

(31.51) $$q \prod_{n \geq 1} \frac{(1 - q^{2n})^4}{(1 - q^{2n-1})^4} = \sum_{N \geq 1} \sigma(2N - 1)q^N,$$

or, on replacing q by q^2 and multiplying by q^{-1},

(31.52) $$q \prod_{n \geq 1} \frac{(1 - q^{4n})^4}{(1 - q^{4n-2})^4} = \left\{ \sum_{k \geq 0} q^{(k+1/2)^2} \right\}^4 = \sum_{N \text{ odd}} \sigma(N)q^N,$$

which is equivalent to the arithmetic result, for odd N,

(31.53) $$\sum 1 \ (4N = \omega_1^2 + \omega_2^2 + \omega_3^2 + \omega_4^2; \omega_i \text{ odd} > 0) = \sigma(N).$$

Similarly, for $p = 4$, $r = 1$,

(31.54) $$q \left\{ \prod_{n \geq 1} \frac{(1 - q^{4n})^2}{(1 - q^{2n-1})} \right\}^2 = \frac{1}{4} \sum_{N \geq 1} \sigma(4N - 1)q^N,$$

or, on replacing q by q^8 and multiplying by q^{-2}, then equating coefficients,

$$(31.55) \quad \sum 1 \ (8N - 2 = \omega_1^2 + \omega_2^2 + 2\omega_3^2 + 2\omega_4^2; \ \omega_i \text{ odd } > 0) = \frac{1}{4}\sigma(4N - 1).$$

As a final example, divide (18.86) by $1 + t$ and let $t \to -1$. We find

$$(31.6) \quad \prod_{n \geq 1} \left(\frac{1 - q^n}{1 + q^n} \right)^8 = 1 + 16 \sum_{N \geq 1} q^N \sum_{d \mid N} (-1)^d d^3,$$

from which we get the theorem

$$(31.61) \quad \sum 1 \ (N = a_1^2 + \cdots + a_8^2) = 16(-1)^N \sum_{d \mid N} (-1)^d d^3.$$

32. Products with multiplicative series.

By a *multiplicative series* we mean one of the form

$$(32.1) \qquad A + B \sum_{N=1}^{\infty} C(N)q^N,$$

where $C(MN) = C(M)C(N)$ for $(M, N) = 1$. Such series which represent modular functions of τ $(q = \exp 2\pi i \tau)$ have been studied by Hecke [26], who has shown a deep-lying connection between the arithmetic properties of the co-efficients and the group- and function-theoretic behavior of the corresponding modular functions. Our purpose here is to exhibit a number of such series for which the function has the form

$$(32.11) \qquad q^b f^{c_1}(q^{a_1}) f^{c_2}(q^{a_2}) \cdots f^{c_r}(q^{a_r}),$$

where the a_i and c_i are integers, $a_i > 0$, $b = 0$ or 1, and

$$(32.12) \qquad f(q) = \prod_{n \geq 1} (1 - q^n).$$

The function $f(q)$ is essentially equal to Dedekind's

$$(32.13) \qquad \eta(\tau) = e^{\pi i \tau / 12} \prod_{n \geq 1} (1 - e^{2\pi i n \tau}),$$

with $q = \exp 2\pi i \tau$. The products (32.11) will, in addition, satisfy the condition that the "weight" $w = c_1 a_1 + c_2 a_2 + \cdots + c_r a_r$ has the value $24b$, so they may be written as products of η-functions. In this section we shall use the abbreviation

$$(32.14) \qquad [a] = \eta(a\tau),$$

so that (32.11) appears as

$$(32.15) \qquad [a_1]^{c_1} [a_2]^{c_2} \cdots [a_r]^{c_r}.$$

Ramanujan [23] has given a number of examples of this type. For instance, he gives

$$(32.2) \qquad [1]^{24} = qf^{24}(q) = \sum_{N \geq 1} \tau(N)q^N.$$

The coefficients $\tau(N)$ were proved to be multiplicative by Mordell [31]. A trivial example is

$$\prod_{n\geq 1}(1 - q^{2n})(1 + q^{2n-1})^2 = 1 + 2\sum_{N\geq 1} e(N)q^N,$$

where $e(N) = 1$ or 0 according as N is a square or not. Since
(32.21)

$$\prod(1 + q^{2n-1}) = \prod\frac{(1 + q^n)}{(1 + q^{2n})} = \prod\left(\frac{1 - q^{2n}}{1 - q^n}\right)\left(\frac{1 - q^{2n}}{1 - q^{4n}}\right) = \frac{f^2(q^2)}{f(q)f(q^4)},$$

we have

(32.22) $$\frac{[2]^5}{[1]^2[4]^2} = 1 + 2\sum_{N\geq 1} e(N)q^N.$$

In the future we shall omit elementary reductions like (32.21); also $\sum$ will, unless otherwise stated, indicate a summation over all positive integers N.

Other examples that we have already found here are the following:

(32.23) $$\frac{[2]^{10}}{[1]^4[4]^4} = 1 + 4\sum E_1(N;4)q^N \quad \text{from (26.63)}.$$

(32.24) $$\frac{[2]^3[4]^3}{[1]^2[8]^2} = 1 + 2\sum E_{1,3}(N;8)q^N \quad \text{from (31.12)}.$$

(32.25) $$\frac{[2]^5[6]^5}{[1]^2[3]^2[4]^2[12]^2} = 1 + 2\sum u(N)q^N \quad \text{from (31.17)}.$$

Equation (31.2) may be written

$$\frac{f^4(q^2)}{f^2(q)} = \sum E_1(4N + 1;4)q^N.$$

Both the form of the coefficients and the weight of the product suggest that we replace q by q^4 and multiply by q, to get

(32.26) $$\frac{[8]^4}{[4]^2} = \sum_{N \text{ odd}} E_1(N;4)q^N,$$

where we have used the fact that $E_1(4k - 1;4) = 0$. Since $(1 - (-1)^N)/2$ is multiplicative as well as $E_1(N;4)$, their product, the coefficient of q^N in (32.26), is also. Similarly, from (31.22), we obtain

(32.27) $$\frac{[4]^2[12]^2}{[2][6]} = \sum_{N \text{ odd}} E_1(N;3)q^N.$$

The four-square theorem (31.31) is equivalent to

(32.28) $$\frac{[2]^{20}}{[1]^8[4]^8} = 1 + 8\sum q^N(2 + (-1)^N)\sum_{\substack{\omega \mid N \\ \omega \text{ odd}}} \omega.$$

Equation (31.32) yields

(32.29) $$\frac{[2]^6[4]^6}{[1]^4[8]^4} = 1 + 4\sum q^N k(\alpha)\sigma(m),$$

where $N = 2^\alpha m$, m odd, and $k(0) = 1$, $k(1) = 2$, $k(\alpha) = 6$ for $\alpha \geq 2$. From (31.43), we have

$$(32.3) \qquad \frac{[2]^{10}[6]^{10}}{[1]^4[3]^4[4]^4[12]^4} = 1 + 4\sum q^N(-1)^{N-1}\sum_{d\,|\,N} d\chi(d),$$

where $\chi(d)$ is defined by (31.42). From (31.52),

$$(32.31) \qquad \frac{[4]^8}{[2]^4} = \sum_{N\text{ odd}} \sigma(N)q^N.$$

The eight-squares theorem (31.61) gives us

$$(32.32) \qquad \frac{[2]^{40}}{[1]^{16}[4]^{16}} = 1 + 16\sum q^N(-1)^{N-1}\sum_{d\,|\,N}(-1)^{d-1}d^3.$$

Let us put $u = 2\pi/3$ in (9.3). Then

$$\prod_{n\geq 1}\frac{(1-q^n)^3}{(1-q^{3n})} = 1 - 3\sum \lambda(N)q^N,$$

where

$$\lambda(N) = \sum_{\substack{\omega\,|\,N \\ \omega\text{ odd}}} \chi\left(\frac{2N}{\omega} - \omega\right),$$

and

$$\chi(k) = \begin{cases} 1, & k \equiv 1 \ (\mathrm{mod}\,6), \\ -1, & k \equiv -1 \ (\mathrm{mod}\,6), \\ 0, & \text{otherwise.} \end{cases}$$

After some analysis, it turns out that $\lambda(N)$ is a multiplicative function, and, in fact

$$(32.33) \qquad \lambda(N) = E_1(N;3) - 3E_1(N/3;3),$$

with the usual convention that $E_1(N/3;3) = 0$ unless $N \equiv 0 \ (\mathrm{mod}\,3)$. Thus we have

$$(32.34) \qquad \frac{[1]^3}{[3]} = 1 - 3\sum\{E_1(N;3) - 3E_1(N/3;3)\}q^N.$$

In (31.19), take $p = 3$, $r = s = 1$. Then

$$(32.35) \qquad \prod_{n\geq 1}\frac{(1-q^{3n})^3}{(1-q^n)} = \sum_{n\geq 0} E_1(3n+1;3)q^n.$$

Replacing q by q^3 and multiplying by q, we get

$$q\prod_{n\geq 1}\frac{(1-q^{9n})^3}{(1-q^{3n})} = \sum_{(N,3)=1} E_1(N;3)q^N.$$

Thus we have

$$(32.351) \qquad \frac{[9]^3}{[3]} = \sum_{(N,3)=1} E_1(N;3)q^N.$$

In (18.61), put $u = v = \pi/3$ to get

(32.36)
$$\frac{[2]^2[3]^3}{[1][6]^2} = 1 + \sum h(N)q^N,$$

where

$$h(N) = \sum_{d \mid N} \frac{\sin(d + N/d)\pi/3}{\sin \pi/3}.$$

After applying the usual analysis, considering N in the form $2^a 3^b m$, where $(m,6) = 1$, we find that $h(N)$ is a multiplicative function, defined by

(32.37)
$$\begin{aligned}
h(2^{2\alpha}) &= -3, \quad h(2^{2\alpha-1}) = 0, \quad h(3^\alpha) = -2, \\
h(p^{\alpha+1}) &= \alpha + 1, \qquad p \text{ a prime}, \equiv 1 \pmod{6}, \\
h(q^{2\alpha}) &= 1, \qquad q \text{ a prime}, \equiv -1 \pmod{6}, \\
h(q^{2\alpha-1}) &= 0, \qquad q \text{ a prime}, \equiv -1 \pmod{6},
\end{aligned}$$

where $\alpha > 0$ throughout.

Again, in (19.4), take $k = \alpha = 1$, $p = 6$, $r = s = 1$. Then

$$\prod_{n \geq 1} \frac{(1 - q^{2n})^3 (1 - q^{3n})^2}{(1 - q^n)^2 (1 - q^{6n})} = \sum_{n \geq 0} E_1(6n + 1; 6)q^n.$$

Replacing q by q^6 and multiplying by q, we have

(32.38)
$$\frac{[12]^3[18]^2}{[6]^2[36]} = \sum_{(N,6)=1} E_1(N; 3)q^N.$$

Here we have used the fact that $E_1(6n - 1; 6) = 0$ and that $E_1(N; 6) = E_1(N; 3)$ if $(N, 6) = 1$.

From (31.11), with $p = 3$, $r = 1$, we get the result

(32.39)
$$\frac{[2][3]^6}{[1]^2[6]^3} = 1 + 2 \sum E_{1,2}(N; 6)q^N,$$

and, with $p = 6$, $r = 1$,

(32.4)
$$\frac{[2]^3[3]^2[6]}{[1]^2[4][12]} = 1 + 2 \sum E_{1,5}(N; 12)q^N.$$

In (19.5), let $k = 3$, $\alpha = \exp 2\pi i/3$. The series becomes $1 + \sum A(N)q^N$, where

(32.41) $\quad A(N) = \dfrac{3}{2}E_{r,p-r}(N; 3p) - \dfrac{i\sqrt{3}}{2}(E_{r,2p+r}(N; 3p) + 2E_{2p-r}(N; 3p)).$

For $p = 2$, $r = 1$, we get

(32.42)
$$\frac{[2]^6[3]}{[1]^3[6]^2} = 1 + 3 \sum E_1(N; 6)q^N.$$

Next, for $k = 4$, $\alpha = i$, the product in (19.5) becomes $P = Q(G - iH)$, where

(32.43)
$$Q = \prod_{n \geq 1} \frac{(1 - q^{pn})}{(1 + q^{2pn})(1 - q^{pn-r})(1 - q^{pn-p+r})},$$

$$G - iH = \prod_{n \geq 1}(1 - q^{pn})(1 + iq^{pn-r})(1 - iq^{pn-p+r})$$

(32.44)

$$= \sum_{n=-\infty}^{+\infty} i^n q^{\frac{1}{2}(pn^2 + (p-2r)n)},$$

by Jacobi's product. Hence

(32.45)

$$G = \sum_{-\infty}^{+\infty}(-1)^n q^{2pn^2 + (p-2r)n}$$

$$= \prod_{n \geq 1}(1 - q^{4pn})(1 - q^{4pn-p-2r})(1 - q^{4pn-3p+2r}),$$

and

(32.46)

$$H = q^r \sum_{-\infty}^{+\infty}(-1)^n q^{2pn^2 + (p+2r)n}$$

$$= q^r \prod_{n \geq 1}(1 - q^{4pn})(1 - q^{4pn-p+2r})(1 - q^{4pn-3p-2r}),$$

again by Jacobi's product. The coefficient in the series is
(32.47)
$$A(N) = (1 - i)\{E_r(N; 4p) + iE_{p+r}(N; 4p) - E_{2p+r}(N; 4p) - iE_{3p+r}(N; 4p)\}$$

$$= E_{r,p-r,p+r,2p-r}(N; 4p) - iE_{r,2p-r,3p-r,3p+r}(N; 4p).$$

The real and imaginary parts are then

(32.48)

$$QG = \prod_{n \geq 1} \frac{(1 - q^{pn})(1 - q^{2pn})(1 - q^{4pn-p-2r})(1 - q^{4pn-3p+2r})}{(1 - q^{pn-r})(1 - q^{pn-p+r})}$$

$$= 1 + \sum q^N E_{r,p-r,p+r,2p-r}(N; 4p),$$

(32.49)

$$QH = q^r \prod_{n \geq 1} \frac{(1 - q^{pn})(1 - q^{2pn})(1 - q^{4pn-p+2r})(1 - q^{4pn-3p-2r})}{(1 - q^{pn-r})(1 - q^{pn-p+r})}$$

$$= \sum q^N E_{r,2p-r,3p-r,3p+r}(N; 4p).$$

From these we get two new multiplicative series, with $p = 6$, $r = 1$:

(32.5)

$$\frac{[2][3][8][12]}{[1][24]} = 1 + \sum E_{1,5,7,11}(N; 24)q^N,$$

(32.51)

$$\frac{[2][3][4][24]}{[1][8]} = \sum E_{1,11,17,19}(N; 24)q^N.$$

Next, taking $k = 6$, $a = 1$ in (19.5), we get

(32.52)

$$A(N) = \tfrac{1}{2}\{E_{r,p-r,2p+r,3p-r}(N; 6p) + 2E_{p+r,2p-r}(N; 6p)\}$$

$$- \frac{i\sqrt{3}}{2}E_{r,3p-r,4p-r,5p+r}(N; 6p),$$

and the only case of immediate interest is $p = 2$, $r = 1$, which simplifies to $E_{1,5}(N; 12) + 2E_3(N; 12)$. After the usual transformations of the product, we have

$$(32.53) \qquad \frac{[2][4]^2[6]^3}{[1][3][12]^2} = 1 + \sum q^N \{E_{1,5}(N; 12) + 2E_3(N; 12)\}.$$

The next four examples are similar in nature to (32.26) and (32.27), where the multiplicativity of a function $b(N)$ implies that of $((1 - (-1)^N)/2)b(N)$. For this purpose, we choose $p = 2s$ in (19.4). Since $(r, p) = 1$ and therefore $(r, s) = 1$, every divisor of $pN + rs = s(2N + r)$ that is congruent to $\pm r \pmod{p}$ will also be a divisor of $2N + r$, so that $E_{mp+r}(pN + rs; kp) = E_{mp+r}(2N + r; kp)$. Taking $r = 1$ leads to the argument $2N + 1$, which is what we want. Thus, choosing $k = 2$, $\alpha = -1$, we have

$$(32.54) \qquad \prod_{n \geq 1} \frac{(1 - q^{2sn})^2(1 + q^{2sn-s-1})(1 + q^{2sn-s+1})}{(1 - q^{2sn-1})(1 - q^{2sn-2s+1})(1 + q^{s(2n-1)})^2}$$
$$= \sum_{N \geq 0} q^N E_{1,2s-1}(2N + 1; 4s).$$

For $s = 1$, we recover (32.26). For $s = 2$, after replacing q by q^2 and multiplying by q, we get

$$(32.55) \qquad \frac{[4]^5[16]^2}{[2]^2[8]^3} = \sum_{N \text{ odd}} E_{1,3}(N; 8)q^N.$$

For $s = 3$, by the same method,

$$(32.56) \qquad \frac{[6]^3[8][24]}{[2][12]^2} = \sum_{N \text{ odd}} E_{1,5}(N; 12)q^N.$$

Next, choose $k = 4$, $\alpha = i$, $r = 1$, $s = 3$ in (19.4). The product becomes $Q(U + iV)$, where

$$Q = \prod_{n \geq 1} \frac{(1 - q^{6n})}{(1 - q^{6n-1})(1 - q^{6n-5})(1 + q^{12n-6})},$$

$$U + iV = \prod_{n \geq 1}(1 - q^{6n})(1 + (iq^{-1})q^{3(2n-1)})(1 + (iq^{-1})^{-1}q^{3(2n-1)})$$

$$= \sum_{-\infty}^{+\infty} i^n q^{3n^2+n} = \sum_{n \text{ even}} + \sum_{n \text{ odd}}$$

$$= \sum_{-\infty}^{+\infty}(-1)^n q^{12n^2+2n} + i \sum_{-\infty}^{+\infty}(-1)^n q^{12n^2+10n+2}.$$

Hence, if we define $\phi(q) = U + V$, $\psi(q) = U - V$, we have

$$\varphi(iq) = \sum_{-\infty}^{+\infty} q^{12n^2+2n} - \sum_{-\infty}^{+\infty} q^{12n^2+10n+2}$$

$$= \sum_{-\infty}^{+\infty}(-1)^n q^{3n^2+n} = \prod_{n \geq 1}(1 - q^{2n}),$$

$$\psi(iq) = \sum_{-\infty}^{+\infty} q^{12n^2+2n} + \sum_{-\infty}^{+\infty} q^{12n^2+10n+2}$$

$$= \sum_{-\infty}^{+\infty} q^{3n^2+n} = \prod_{n\geq 1}(1-q^{6n})(1+q^{6n-2})(1+q^{6n-4})$$

$$= \prod_{n\geq 1} \frac{(1-q^{6n})(1+q^{2n})}{(1+q^{6n})}.$$

It follows that

$$Q\phi(q) = Q(U+V) = \prod_{n\geq 1} \frac{(1-q^{3n})(1-q^{4n})^3(1-q^{6n})(1-q^{24n})}{(1-q^n)(1-q^{8n})(1-q^{12n})^2},$$

$$Q\psi(q) = Q(U-V) = \prod_{n\geq 1} \frac{(1-q^{2n})^2(1-q^{3n})(1-q^{8n})(1-q^{12n})^3}{(1-q^n)(1-q^{4n})^2(1-q^{6n})(1-q^{24n})}.$$

Now the real part of the series is

$$QU = \sum_{N\geq 0} q^N\{E_1(2N+1;24) - E_{13}(2N+1;24)\},$$

and the imaginary part is

$$QV = \sum_{N\geq 0} q^N\{E_7(2N+1;24) - E_{19}(2N+1;24)\}.$$

Thus, performing the indicated operations, replacing q by q^2, and multiplying by q, we obtain

(32.57) $$\frac{[6][8]^3[12][48]}{[2][16][24]^2} = \sum_{N \text{ odd}} E_{1,5,7,11}(N;24)q^N,$$

(32.58) $$\frac{[4]^2[6][16][24]^3}{[2][8]^2[12][48]} = \sum_{N \text{ odd}} E_{1,11,17,19}(N;24)q^N.$$

In (19.61), $a=1$, $k=2,3,4$ lead immediately to (32.26), (32.27), and (32.55) respectively. For $k=6$ we get the new series

(32.59) $$\frac{[4]^7[6][24]^2}{[2]^3[8]^2[12]^3} = \sum_{N \text{ odd}} q^N\{E_{1,5}(N;12) + 2E_3(N;12)\}.$$

In (30.1) and (30.11), divide by $(1+a)$ and let $a \to -1$, to get

(32.6) $$\sum_{-\infty}^{+\infty}(6m+1)q^{(3m^2+m)/2} = \prod_{n\geq 1} \frac{(1-q^n)^3}{(1+q^n)^2}.$$

Replacing q by q^{24} and multiplying by q, we have

(32.61) $$\frac{[24]^5}{[48]^2} = \sum \chi(N)q^N,$$

where

$$(32.62) \qquad \chi(N) = \begin{cases} 0 & \text{if } N \neq k^2, \ (k,6) = 1, \\ k & \text{if } N = k^2, \ k \equiv 1 \ (\text{mod } 6), \ k > 0, \\ -k & \text{if } N = k^2, \ k \equiv -1 \ (\text{mod } 6), \ k > 0. \end{cases}$$

Now it is easy to see by Jacobi's identity that

$$L(u) \equiv \sum_{-\infty}^{+\infty} (-1)^m \cos(6m+1)u \, q^{(3m^2+m)/2} = L_1(u) + \overline{L_1(u)},$$

where

$$L_1(u) = \frac{e^{iu}}{2} \prod_{n \geq 1} (1 - q^{3n})(1 - e^{6iu} q^{3n-1})(1 - e^{-6iu} q^{3n-2}).$$

Divide by $\cos u$ and let $u \to \pi/2$. Then

$$\sum_{-\infty}^{+\infty} (6m+1) q^{(3m^2+m)/2} = \lim_{u \to \pi/2} \frac{L(u)}{\cos u} = -L'(\pi/2).$$

But if we differentiate logarithmically, we find

$$L_1'(\pi/2) = i L_1(\pi/2)\{1 - 6g(q)\},$$

where

$$\begin{aligned} g(q) &= \sum_{n \geq 1} \left\{ \frac{q^{3n-2}}{1 + q^{3n-2}} - \frac{q^{3n-1}}{1 + q^{3n-1}} \right\} \\ &= \sum_{n \geq 1} \left\{ \frac{q^{3n-2}}{1 - q^{3n-2}} - \frac{q^{3n-1}}{1 - q^{3n-1}} - 2 \left(\frac{q^{6n-4}}{1 - q^{6n-4}} - \frac{q^{6n-2}}{1 - q^{6n-2}} \right) \right\} \\ &= \sum_{N \geq 1} q^N \{E_1(N;3) - 2E_1(N/2;3)\}. \end{aligned}$$

Since $L'(\pi/2) = 2L_1'(\pi/2)$ and

$$L_1(\pi/2) = \frac{i}{2} \prod_{n \geq 1} \frac{(1 - q^{3n})(1 + q^n)}{(1 + q^{3n})},$$

$$(32.63) \qquad \sum_{-\infty}^{+\infty} (6m+1) q^{(3m^2+m)/2} = \prod_{n \geq 1} \frac{(1 - q^{3n})(1 + q^n)}{(1 + q^{3n})} \{1 - 6g(q)\}.$$

Comparison with (32.6) then yields

$$(32.64) \qquad \frac{[1]^6 [6]}{[2]^3 [3]^2} = 1 - 6 \sum q^N \{E_1(N;3) - 2E_1(N/2;3)\}.$$

Our next set of examples comes from (18.85) and (18.87). In the first, $p = 6$, $s = 1$, $a = 2$ yields

$$(32.65) \qquad \frac{[1]^3 [2]^3}{[3][6]} = 1 - 3 \sum q^N \sum_{d \mid N} d\chi(d),$$

where $\chi(d) - 1$, $d \equiv \pm 1 \pmod 6$; $= -2$, $d \equiv 3 \pmod 6$; $= 0$, otherwise. For $p = 6$, $s = 1$, $a = 3$,

$$(32.66) \qquad \frac{[1]^4[3]^4}{[2]^2[6]^2} = 1 - 4 \sum q^N \sum_{d \mid N} d\chi(d),$$

where $\chi(d) = 1$, $d \equiv \pm 1 \pmod 6$; $= -1$, $d \equiv \pm 2 \pmod 6$; $= 0$, otherwise. For $p = 8$, $s = 1$, $a = 3$,

$$(32.67) \qquad \frac{[1]^2[2][4]^3}{[8]^2} = 1 - 2 \sum q^N \sum_{d \mid N} d\chi(d),$$

where $\chi(d) = 1$, $d \equiv \pm 1 \pmod 8$; $= -1$, $d \equiv \pm 3 \pmod 8$; $= 0$, otherwise. For $p = 12$, $s = 1$, $a = 5$,

$$(32.68) \qquad \frac{[1][3][4]^2[6]^2}{[12]^2} = 1 - \sum q^N \sum_{d \mid N} d\chi(d),$$

where $\chi(d) = 1$, $d \equiv \pm 1 \pmod{12}$; $= -1$, $d \equiv \pm 5 \pmod{12}$; $= 0$, otherwise.

In (18.87), $p = 3$, $a = 1$ yields

$$(32.69) \qquad \frac{[1]^9}{[3]^3} = 1 - 9 \sum q^N \sum_{d \mid N} d^2\chi(d),$$

where $\chi(d) = 1$, $d \equiv 1 \pmod 3$; $= -1$, $d \equiv -1 \pmod 3$; $= 0$, $d \equiv 0 \pmod 3$. From $p = 4$, $a = 1$,

$$(32.7) \qquad \frac{[1]^4[2]^6}{[4]^4} = 1 - 4 \sum q^N \sum_{d \mid N} d^2\chi(d),$$

where $\chi(d) = 1$, $d \equiv 1 \pmod 4$; $= -1$, $d \equiv -1 \pmod 4$; $= 0$, otherwise. From $p = 6$, $a = 1$,

$$(32.71) \qquad \frac{[1][2]^4[3]^5}{[6]^4} = 1 - \sum q^N \sum_{d \mid N} d^2\chi(d),$$

where $\chi(d) = 1$, $d \equiv 1$ or $2 \pmod 6$; $= -1$, $d \equiv -1$ or $-2 \pmod 6$; $= 0$, otherwise.

As a final example, let us put $p = 6$, $r = 1$ in (31.15). Then

$$(32.72) \qquad \frac{[1]^2[4][6]^7}{[2]^3[3]^2[12]^3} = 1 - 2 \sum q^N \{E_{1,5}(N; 12) - 2E_{1,5}(N/2; 12)\}.$$

There are undoubtedly many other examples to be obtained from (18.83) and (18.86) by suitable specialization. It should also be observed that replacing q by $-q$ in any of our examples leads to a new one, except when the function involved is odd. For instance, from (32.22) we derive

$$(32.73) \qquad \frac{[1]^2}{[2]} = 1 - 2 \sum (-1)^{N-1} e(N) q^N.$$

It should be noticed, too, that the identities of Euler, Gauss, and Jacobi, for

$$\prod_{n \geq 1} (1 - q^n), \quad \prod_{n \geq 1} \frac{(1 - q^{2n})}{(1 - q^{2n-1})}, \quad \prod_{n \geq 1} (1 - q^n)^3,$$

respectively, lead to series of the type we have been considering.

33. More about products. Equation (32.65) may be written as

$$\frac{\eta^3(\tau)\eta^3(2\tau)}{\eta(3\tau)\eta(6\tau)} = 1 - 3\sum_{k\geq 1}\frac{k\chi(k)q^k}{1-q^k}$$

(33.1)
$$= 1 + q\frac{d}{dq}\log\prod_{k\geq 1}(1-q^k)^{3\chi(k)}$$

$$= \frac{(qG(q))'}{G(q)},$$

where

$$G(q) = \prod(1-q^k)^{3\chi(k)} = \prod\frac{(1-q^k)^3(1-q^{6k})^9}{(1-q^{2k})^3(1-q^{3k})^9},$$

(33.11)
$$qG(q) = \frac{\eta^3(\tau)\eta^9(6\tau)}{\eta^3(2\tau)\eta^9(3\tau)} = W(\tau).$$

Since $dW(\tau)/dq = (1/2\pi iq)W'(\tau)$, (33.1) yields

(33.12)
$$W'(\tau) = 2\pi i\frac{\eta^6(\tau)\eta^8(6\tau)}{\eta^{10}(3\tau)}.$$

From the theory of the η-function we recall the fundamental transformation equation

(33.2)
$$\eta(-1/\tau) = (-i\tau)^{1/2}\eta(\tau),$$

using that branch of the square-root with positive real part. Applying (33.2) in (33.12), we get

$$\frac{d}{d\tau}W\left(-\frac{1}{6\tau}\right) = W'\left(-\frac{1}{6\tau}\right)\cdot\frac{1}{6\tau^2} = \frac{2\pi i}{6\tau^2}\cdot\frac{\eta^6(-1/6\tau)\eta^8(-1/\tau)}{\eta^{10}(-1/2\tau)}$$

(33.121)
$$= -2\pi i\cdot\frac{9}{8}\cdot\frac{\eta^8(\tau)\eta^6(6\tau)}{\eta^{10}(2\tau)}.$$

Now, again by (33.2),

$$W(-1/6\tau) = \frac{1}{8}\cdot\frac{\eta^9(\tau)\eta^3(6\tau)}{\eta^9(2\tau)\eta^3(3\tau)}.$$

Hence

(33.122)
$$\frac{d}{d\tau}\left(\frac{\eta^9(\tau)\eta^3(6\tau)}{\eta^9(2\tau)\eta^3(3\tau)}\right) = 2\pi i(-9)\cdot\frac{\eta^8(\tau)\eta^6(6\tau)}{\eta^{10}(2\tau)},$$

(33.123)
$$\frac{d}{dq}\prod\frac{(1-q^n)^9(1-q^{6n})^3}{(1-q^{2n})^9(1-q^{3n})^3} = -9\prod\frac{(1-q^n)^8(1-q^{6n})^6}{(1-q^{2n})^{10}}.$$

After routine calculations, this reduces to

$$\frac{\eta^3(3\tau)\eta^3(6\tau)}{\eta(\tau)\eta(2\tau)} = q\prod\frac{(1-q^{3n})^3(1-q^{6n})^3}{(1-q^n)(1-q^{2n})}$$

(33.124)
$$= \sum_{k\geq 1}\frac{k\chi_1(k)q^k}{1-q^k},$$

where χ_1 is the multiplicative function defined by $\chi_1(k) = 1$, $k \equiv \pm 1 \pmod 6$; $= 2/3$, $k \equiv 3 \pmod 6$; $= 0$, else. The series in (33.124) can of course be written as

$$(33.125) \qquad \sum_{N \geq 1} q^N \sum_{d \mid N} d\chi_1(d).$$

Similarly, from (32.66), by the same process, we get

$$(33.2) \qquad \begin{aligned} \frac{\eta^4(2\tau)\eta^4(6\tau)}{\eta^2(\tau)\eta^2(3\tau)} &= \sum \frac{k\chi_2(k)q^k}{1 - q^k} \\ &= \sum q^N \sum_{d \mid N} d\chi_2(d), \end{aligned}$$

where χ_2 is the multiplicative function defined by $\chi_2(k) = 1$, $k \equiv \pm 1, \pmod 6$; $= \frac{1}{2}$, $k \equiv \pm 2 \pmod 6$; $= 0$, else.

In (18.85), put $p = 6$, $a = 3$, $s = 2$. This yields

$$(33.3) \qquad \frac{\eta^{12}(\tau)\eta^2(6\tau)}{\eta^6(2\tau)\eta^4(3\tau)} = 1 - 12 \sum \frac{k\chi(k)q^k}{1 - q^k},$$

where χ (not multiplicative) is defined by $\chi(k) = 1$, $k \equiv \pm 1 \pmod 6$; $= -3$, $k \equiv \pm 2 \pmod 6$; $= 4$, $k \equiv 3 \pmod 6$; $= 0$, $k \equiv 0 \pmod 6$.

Now, from (32.65), (32.66), (33.3), it is clear that some linear combination of the functions involved is a constant. By comparing the first three coefficients in either the products or the series, we see that

$$(33.31) \qquad 8 \cdot \frac{[1]^3[2]^3}{[3][6]} - 9 \cdot \frac{[1]^4[3]^4}{[2]^2[6]^2} + \frac{[1]^{12}[6]^2}{[2]^6[3]^4} = 0.$$

If we subject (33.31) to the transformation $\tau \to -1/6\tau$ and use (33.2), we get

$$(33.32) \qquad \frac{[3]^3[6]^3}{[1][2]} - \frac{[2]^4[6]^4}{[1]^2[3]^2} + \frac{[1]^2[6]^{12}}{[2]^4[3]^6} = 0.$$

Let $u = u(\tau) = \eta(2\tau)/\eta(\tau) = [2]/[1]$, $v = v(\tau) = u(3\tau) = \eta(6\tau)/\eta(3\tau) = [6]/[3]$. Then (33.31) becomes

$$(33.33) \qquad 9\left(\frac{[3]}{[1]}\right)^4 = 8u^5 v + \frac{v^4}{u^4},$$

and (33.32) becomes

$$(33.34) \qquad u^8 v = \left(\frac{[3]}{[1]}\right)^4 (u^3 + v^9).$$

Eliminating $[3]/[1]$ between (33.33) and (33.34), we get

$$(33.35) \qquad 9 = (8 + (v/u^3)^3)(1 + (v^3/u)^3).$$

Define $h(\tau) = (v/u^3)^3$. Using (33.2), we find that

$$h(-1/6\tau) = 8(v^3/u)^3,$$

so (33.35) becomes

$$(33.351) \qquad 72 = (8 + h(\tau))(8 + h(-1/6\tau)).$$

Now if $\tau = \tau_0 = i/\sqrt{6}$, we have

$$72 = (8 + h(\tau_0))^2,$$
$$h(\tau_0) = -8 \pm 6\sqrt{2}.$$

From its definition we see that $h(\tau_0) > 0$, so

(33.352) $$h(\tau_0) = 6\sqrt{2} - 8.$$

Again, from (33.2),

$$u(-1/6\tau) = 1/\sqrt{2}v(\tau),$$

so $u(\tau_0)v(\tau_0) = 1/\sqrt{2}$. Combining this with (33.352), we find

(33.353) $$u\left(\frac{i}{\sqrt{6}}\right) = u(\tau_0) = \left(\frac{3 + 2\sqrt{2}}{8}\right)^{1/12},$$

that is,

(33.354) $$e^{-\pi/12\sqrt{6}} \prod_{n \geq 1}(1 + e^{-2\pi n/\sqrt{6}}) = \left(\frac{3 + 2\sqrt{2}}{8}\right)^{1/12}.$$

If we put these known values of $u(\tau_0)$ and $v(\tau_0)$ back in (33.33), we can solve for $[3]/[1]$:

(33.355) $$\frac{\eta(3\tau_0)}{\eta(\tau_0)} = \left(\frac{3 + 2\sqrt{2}}{27}\right)^{1/12}.$$

The four-square theorem, or its equivalent (32.28), may be written as

(33.4) $$\frac{\eta^{20}(2\tau)}{\eta^8(\tau)\eta^8(4\tau)} = 1 + 8\sum \frac{k\alpha(k)q^k}{1 - q^k},$$

where $\alpha(k) = 0$ if $k \equiv 0 \pmod 4$, $= 1$ if $k \not\equiv 0 \pmod 4$. As in (33.1) et seq., this becomes

(33.41) $$\prod \frac{(1 - q^{2n})^{20}}{(1 - q^n)^8(1 - q^{4n})^8} = q\frac{d}{dq}\log\left(q\prod\frac{(1 - q^{4n})^8}{(1 - q^n)^8}\right).$$

Hence

(33.411) $$\frac{d}{dq}\left\{q\prod\frac{(1 - q^{4n})^8}{(1 - q^n)^8}\right\} = \prod\frac{(1 - q^{2n})^{20}}{(1 - q^n)^{16}},$$

or, for real q, $|q| < 1$,

(33.42) $$q\prod\frac{(1 - q^{4n})^8}{(1 - q^n)^8} = \int_0^q \prod\frac{(1 - q^{2n})^{20}}{(1 - q^n)^{16}}dq.$$

The left side of (33.42) is $(\eta(4\tau)/\eta(\tau))^8$. If $\tau = i/2$, $4\tau = -1/\tau$, so by (33.2),

$$\left(\frac{\eta(4\tau)}{\eta(\tau)}\right)^8 = \left(\frac{\eta(-1/\tau)}{\eta(\tau)}\right)^8 = (-i\tau)^4 = \frac{1}{16}.$$

Also $q = e^{2\pi i\tau} = e^{-\pi}$. Putting these values in (33.42), we get

(33.43)
$$\int_0^{e^{-\pi}} \prod_{n\geq 1} \frac{(1-q^{2n})^{20}}{(1-q^n)^{16}} dq = \frac{1}{16}.$$

Similarly, the representation theorem for $a^2 + b^2 + 2c^2 + 2d^2$, (31.32), may be written as

(33.5)
$$\left(\sum_{-\infty}^{+\infty} q^{n^2}\right)^2 \left(\sum_{-\infty}^{+\infty} q^{2n^2}\right)^2 = 1 + 4\sum_{k\geq 1} \frac{k\beta(k)q^k}{1-q^k},$$

where $\beta(k) = 1$, k odd;$= \frac{1}{2}$, $k \equiv 2 \pmod 4$;$= 1$, $k \equiv 4 \pmod 8$;$= 0$, $k \equiv 0 \pmod 8$. (The more familiar form is, for m odd, $\alpha \geq 0$,

$$R(2^\alpha m) = \begin{cases} 4\sigma(m) & (\alpha = 0), \\ 8\sigma(m) & (\alpha = 1), \\ 24\sigma(m) & (\alpha \geq 2).\end{cases}$$

The left side of (33.5), by the Jacobi triple product and some elementary transformations, is

$$\prod \frac{(1-q^{2n})^6(1-q^{4n})^6}{(1-q^n)^4(1-q^{8n})^4}.$$

The right side, as in (33.1), is

$$q\frac{d}{dq}\log(qP) = \frac{1}{P}\frac{d}{dq}(qP),$$

where

$$P = \prod \frac{(1-q^{2n})^2(1-q^{8n})^4}{(1-q^n)^4(1-q^{4n})^2}.$$

Therefore

(33.51)
$$\frac{d}{dq}(qP) = \prod \frac{(1-q^{2n})^8(1-q^{4n})^4}{(1-q^n)^8}.$$

For q real, $|q| < 1$, we have

(33.52)
$$\frac{\eta^2(2\tau)\eta^4(8\tau)}{\eta^2(4\tau)\eta^4(\tau)} = \int_0^q \prod \frac{(1-q^{2n})^8(1-q^{4n})^4}{(1-q^n)^8} dq.$$

Let $\tau = i/2\sqrt{2}$, so that $2\tau = -1/4\tau$, $8\tau = -1/\tau$. The left side of (33.52) is

$$\frac{\eta^2(-1/4\tau)\eta^4(-1/\tau)}{\eta^2(4\tau)\eta^4(\tau)} = (-4i\tau)(-i\tau)^2 = \frac{\sqrt{2}}{8}.$$

Therefore

(33.53)
$$\int_0^{e^{-\pi/\sqrt{2}}} \prod_{n\geq 1} \frac{(1-q^{2n})^8(1-q^{4n})^4}{(1-q^n)^8} dq = \frac{\sqrt{2}}{8}.$$

In (31.4), replace q by $-q$ to get the representation theorem for $a^2 + b^2 + 3c^2 + 3d^2$:

(33.6)
$$S = \frac{[2]^{10}[6]^{10}}{[1]^4[3]^4[4]^4[12]^4} = 1 + 4\sum_{N\geq 1} q^N(-1)^{N-1}\sum_{d\mid N} d\chi(d),$$

where χ is defined in (31.42). It is not difficult to see that

$$(-1)^{N-1} \sum_{d \mid N} d\chi(d) = \sum_{d \mid N} d\gamma(d),$$

where $\gamma(n) = 1$, $n \equiv \pm 1, \pm 4, \pm 5 \pmod{12}$; $= 0$, else. Hence

$$S = 1 + 4 \sum_{k \geq 1} \frac{k\gamma(k)q^k}{1 - q^k} = q \frac{d}{dq} \log(qT),$$

where, after routine calculations,

$$T = \prod \frac{(1 - q^{2n})^4 (1 - q^{3n})^4 (1 - q^{12n})^4}{(1 - q^n)^4 (1 - q^{4n})^4 (1 - q^{6n})^4}.$$

Therefore

(33.61) $$\frac{d}{dq}(qT) = TS = \prod \frac{(1 - q^{2n})^{14}(1 - q^{6n})^6}{(1 - q^n)^8 (1 - q^{4n})^8}.$$

For q real, $|q| < 1$,

(33.62) $$\frac{\eta^4(2\tau)\eta^4(3\tau)\eta^4(12\tau)}{\eta^4(\tau)\eta^4(4\tau)\eta^4(6\tau)} = \int_0^q \prod_{n \geq 1} \frac{(1 - q^{2n})^{14}(1 - q^{6n})^6}{(1 - q^n)^8 (1 - q^{4n})^8} dq.$$

Now let $\tau = i/2\sqrt{3}$, so that $12\tau = -1/\tau$, $3\tau = -1/4\tau$, $2\tau = -1/6\tau$. Then, as above, the left side of (33.62) equals $1/3$, so

(33.63) $$\int_0^{e^{-\pi/\sqrt{3}}} \prod_{n \geq 1} \frac{(1 - q^{2n})^{14}(1 - q^{6n})^6}{(1 - q^n)^8 (1 - q^{4n})^8} dq = \frac{1}{3}.$$

Let us recall

(32.23) $$\frac{[2]^{10}}{[1]^4[4]^4} = 1 + 4 \sum_{N \geq 1} E_1(N; 4)q^N,$$

(32.26)
$$\frac{[8]^4}{[4]^2} = \sum_{N \text{ odd}} E_1(N; 4)q^N$$
$$= \sum_{N \geq 1} E_1(N; 4)q^N - \sum_{N \geq 1} E_1(2N; 4)q^{2N}$$
$$= \sum_{N \geq 1} E_1(N; 4)q^N - \sum_{N \geq 1} E_1(N; 4)q^{2N}$$
$$= \frac{1}{4}\left(\frac{[2]^{10}}{[1]^4[4]^4} - 1\right) - \frac{1}{4}\left(\frac{[4]^{10}}{[2]^4[8]^4} - 1\right),$$

(33.7) $$4\frac{[8]^4}{[4]^2} = \frac{[2]^{10}}{[1]^4[4]^4} - \frac{[4]^{10}}{[2]^4[8]^4}.$$

If we apply the transformation $\tau \to -1/8\tau$, this becomes

(33.71) $$\frac{[1]^4}{[2]^2} = 2 \cdot \frac{[4]^{10}}{[2]^4[8]^4} - \frac{[2]^{10}}{[1]^4[4]^4}.$$

Let $u = [2]/[1]$, $v = [4]/[2]$, $w = [8]/[4]$. Then (33.7) is equivalent to

(33.72)
$$4w^4 = \frac{u^4}{v^6} - \frac{v^4}{w^4},$$

and (33.71) to

(33.73)
$$\frac{1}{u^4} = 2\frac{v^6}{w^4} - \frac{u^4}{v^4}.$$

Elimination of w^4 between these two equations leads to

(33.74)
$$16u^8v^{16} - u^{16} + v^8 = 0.$$

Let $r = r(\tau) = u^8(\tau) = (\eta(2\tau)/\eta(\tau))^8$, $s = s(\tau) = r(2\tau)$. Then (33.74) is an algebraic relation between $r(\tau)$ and $r(2\tau)$, an example of a *modular equation*.

(33.75)
$$16rs^2 - r^2 + s = 0.$$

Let $\tau = i/\sqrt{2}$. Then $2\tau = -1/\tau$, so

$$r = \left(\frac{\eta(2\tau)}{\eta(\tau)}\right)^8 = \left(\frac{\eta(-1/\tau)}{\eta(\tau)}\right)^8 = (-i\tau)^4 = \frac{1}{4}.$$

Using this value in (33.75), we can solve for s:

(33.76)
$$s = \left(\frac{\eta(4\tau)}{\eta(2\tau)}\right)^8 = \left(\frac{\eta(\sqrt{-8})}{\eta(\frac{1}{2}\sqrt{-8})}\right)^8 = \frac{\sqrt{2}-1}{8}.$$

This is equivalent to the evaluation of the class-invariant

(33.77)
$$f_1(\sqrt{-8})^8 = 8(1 + \sqrt{2}),$$

as defined by $f_1(\tau) = \eta(\tau/2)/\eta(\tau)$. See [46].

Notes

The topics in this chapter cover a miscellany of items that naturally arise from the foundations of Chapter 1.

§28. There is a good introduction to theorems of Liouville type in Uspensky and Heaslet's little book on number theory [7]. Also Carlitz [2] provides an interesting account and extension of related work.

§31. The study of sums of squares is treated in depth in the recent book of E. Grosswald [3]. Also in [1] the formulas for 2, 4, and 8 squares are derived in a different manner from q-hypergeometric functions.

§32. The work in this section (as is pointed out) is closely related to the extensive study made by Hecke of modular functions with multiplicative coefficients. R. Rankin [6, §9.5] provides an excellent history of recent work. Readers interested in modern readable extensions of Hecke's ideas should consult A. Ogg [5] or A. Weil [8]. J. Lehner [4] provides an excellent self-contained account.

§33. In this continuation of §22, the author presents a number of amazing identities related to infinite products. The work is quite reminiscent of Ramanujan, especially surprising integrals like (33.43), (33.53), and (33.63).

References

1. G. E. Andrews, *Applications of basic hypergeometric functions*, S.I.A.M. Rev. **16** (1974), 441–484.

2. L. Carlitz, *Bulyzin's method for sums of squares*, J. Number Theory **5** (1973), 405–412.

3. E. Grosswald, *Representations of integers as sums of squares*, Springer-Verlag, 1985.

4. J. Lehner, *Lectures on modular forms*, Nat. Bur. Standards. Appl. Math Series, no. 61, Washington, D.C., 1969.

5. A. Ogg, *Modular forms and Dirichlet series*, Benjamin, New York, 1969.

6. R. Rankin, *Modular forms and functions*, Cambridge Univ. Press, 1977.

7. J. V. Uspensky and M. A. Heaslet, *Elementary number theory*, McGraw-Hill, 1939.

8. A. Weil, *Dirichlet series and automorphic forms*, Lecture Notes in Math., No. 189, Springer-Verlag, 1971.

CHAPTER 5

Modular Equations

34. Modular equations, preliminaries. In this section we introduce some standard definitions from the theory of elliptic functions and elliptic theta-functions, leading up to the topic of modular equations in irrational form. The only fact to be borrowed is the transformation formula for the theta-functions, equation (34.2). We refer the reader, for example, to [**28**].

With $q = e^{\pi i \tau}$, $I(\tau) > 0$, we define

$$(34.1) \qquad \vartheta_3(\tau) = \sum_{-\infty}^{+\infty} q^{n^2} = \prod_{n \geq 1}(1 - q^{2n})(1 + q^{2n-1})^2.$$

Here, as in what follows, we shall be dealing with theta-functions of 0 argument, $\vartheta(0|\tau)$, which will be denoted simply by $\vartheta(\tau)$. To continue,

$$(34.11) \qquad \vartheta_4(\tau) = \vartheta_3(\tau+1) = \sum_{-\infty}^{+\infty}(-1)^n q^{n^2} = \prod_{n \geq 1}\left(\frac{1-q^n}{1+q^n}\right),$$

$$(34.12) \qquad \vartheta_2(\tau) = 2q^{1/4}\prod_{n \geq 1}\left(\frac{1-q^{4n}}{1-q^{4n-2}}\right) = 2\sum_{\omega \text{ odd} > 0} q^{\omega^2/4}.$$

For these functions we have the transformations

$$(34.2) \qquad \begin{aligned} \vartheta_3(-1/\tau) &= \sqrt{-i\tau}\,\vartheta_3(\tau), \\ \vartheta_2(-1/\tau) = \sqrt{-i\tau}\,\vartheta_4(\tau), \qquad \vartheta_4(-1/\tau) &= \sqrt{-i\tau}\,\vartheta_2(\tau), \end{aligned}$$

with that branch of the square-root with positive real part.

Further, we define

$$(34.3) \qquad \sqrt{k} \equiv \sqrt{k(\tau)} = \frac{\vartheta_2(\tau)}{\vartheta_3(\tau)}; \qquad \sqrt{k'} \equiv \sqrt{k'(\tau)} = \frac{\vartheta_4(\tau)}{\vartheta_3(\tau)}.$$

It is clear from (34.2) that

$$(34.31) \qquad \sqrt{k(-1/\tau)} = \sqrt{k'(\tau)}.$$

There is also a simple algebraic relation between k and k':

$$(34.32) \qquad k^2 + k'^2 = 1.$$

To see this, we observe that, by (31.31),

$$\vartheta_3^4(\tau) = 1 + \sum_{N \geq 1} R_4(N)q^N,$$

where $R_4(N) = 8\sigma(N)$ for N odd. Hence (31.52) and (34.12) yield

$$\vartheta_2^4(\tau) = 2 \sum_{N \text{ odd}} R_4(N)q^n.$$

Subtracting, we have

$$\vartheta_3^4(\tau) - \vartheta_2^4(\tau) = 1 + \sum_{N \text{ even}} R_4(N)q^N - \sum_{N \text{ odd}} R_4(N)q^N$$

$$= 1 + \sum_{N \geq 1} (-1)^N R_4(N)q^N = \vartheta_3^4(\tau + 1) = \vartheta_4^4(\tau),$$

which is equivalent to (34.32).

Continuing with our definitions, we let

(34.4) $K \equiv K(\tau) = (\pi/2)\vartheta_3^2(\tau); \qquad K' \equiv K'(\tau) = K(-1/\tau).$

From (34.2) we have the relations

(34.41) $\tau = iK'/K; \qquad q = e^{\pi i \tau} = e^{-\pi K'/K}.$

We shall be concerned with transformations of k, k', K, K' by $\tau \to p\tau$, where p is an arbitrary positive integer. The resulting functions will then be denoted by

(34.5) $\begin{aligned} k(p\tau) &= l(\tau); & k'(p\tau) &= l'(\tau), \\ K(p\tau) &= L(\tau); & K'(p\tau) &= L'(\tau). \end{aligned}$

By definition and (34.2), under $\tau \to -1/p\tau$,

(34. 51) $\begin{aligned} k &\leftrightarrow l'; & k' &\leftrightarrow l, \\ K &\leftrightarrow L'; & K' &\leftrightarrow L. \end{aligned}$

Also, by (34.41),

(34.52) $\dfrac{L'}{L} = p\dfrac{K'}{K}.$

Therefore, if we define $R = K/L$, we see that the transformation $\tau \to -1/p\tau$ has the effect

(34.53) $R \leftrightarrow p/R$

Thus, if by some means we can obtain an equation of the form

(34.6) $R = \varphi(k, k', l, l')$

the transformation $\tau \to -1/p\tau$ will yield, by (34.51) and (34.53), the equation

(34.61) $p/R = \varphi(l', l, k', k).$

This enables us to eliminate R and to obtain a relation between the moduli k, k' and the corresponding moduli l, l'. Such a relation we shall call a modular equation of degree p (in irrational form).

As an example, consider the case $p = 2$. Directly from the products (34.1) and (34.11), it is easy to see that

(34.7) $$\vartheta_4^2(2\tau) = \vartheta_3(\tau)\vartheta_4(\tau),$$

which is equivalent to

(34.71) $$R = K/L = l'/\sqrt{k'},$$

leading to the modular equation of degree 2,

(34.72) $$l'k/\sqrt{k'l} = 2.$$

An alternate method would be to consider the even part of $\vartheta_3^2(\tau)$:

(34.73)
$$\vartheta_3^2(\tau) + \vartheta_4^2(\tau) = 2\left\{1 + 4\sum_{N \text{ even}} E_1(N;4)q^N\right\}$$
$$= 2\left\{1 + 4\sum_{N \geq 1} E_1(2N;4)q^{2N}\right\}$$
$$= 2\vartheta_3^2(2\tau),$$

since $E_1(2N;4) = E_1(N;4)$. Equation (34.73) is equivalent to

(34.74) $$R = 2/(1 + k'),$$

leading to the modular equation

(34.75) $$(1 + k')(1 + l) = 2.$$

On the other hand, if we equate (34.71) and (34.74), we obtain

(34.76) $$1/l' = \tfrac{1}{2}(\sqrt{k'} + 1/\sqrt{k'}).$$

The three forms (34.72), (34.75), and (34.76) are obviously interdependent, for otherwise (34.32) and the corresponding relation for l, l' would lead to a nontrivial algebraic equation for $k(\tau)$. In general, the modular equation may present itself in many different forms, some of striking elegance and simplicity. Another form, for $p = 2$, is

(34.77) $$l = \frac{1 - k'}{1 + k'} = \left(\frac{k}{1 + k'}\right)^2 = \left(\frac{1 - k'}{k}\right)^2.$$

In case τ is chosen so as to be equal to $-1/p\tau$, that is, $\tau = i/\sqrt{p}$, the modular equation will lead to an algebraic equation determining the modulus $k'(i/\sqrt{p})$, since then $l = k'$ and $l' = k$. Thus for $\tau = i/\sqrt{2}$, (34.75) yields

(34.78) $$k'(i/\sqrt{2}) = \sqrt{2} - 1,$$

the positive sign being chosen because k' is positive for pure imaginary τ.

Turning to the case $p = 3$, we select equation (32.27), which may be rewritten

$$4\sum_{N \text{ odd}} E_1(N;6)q^N = 4q\prod_{n \geq 1}\left(\frac{1 - q^{4n}}{1 - q^{4n-2}}\right)\left(\frac{1 - q^{12n}}{1 - q^{12n-6}}\right)$$
$$= \vartheta_2(\tau)\vartheta_2(3\tau).$$

Also, from (31.16),

$$\vartheta_4(\tau)\vartheta_4(3\tau) = 1 - 2\sum_{N\geq 1} q^N\{E_{1,2}(N;6) - 2E_{2,4}(N;12)\},$$

and, replacing q by $-q$,

$$\vartheta_3(\tau)\vartheta_3(3\tau) = 1 - 2\sum_{N\geq 1} (-1)^N q^N\{E_{1,2}(N;6) - 2E_{2,4}(N;12)\}.$$

Comparing coefficients, we find

$$(34.8) \qquad \vartheta_3(\tau)\vartheta_3(3\tau) - \vartheta_4(\tau)\vartheta_4(3\tau) = \vartheta_2(\tau)\vartheta_2(3\tau),$$

which is equivalent to the modular equation of degree 3, in one of the standard forms,

$$(34.81) \qquad\qquad \sqrt{kl} + \sqrt{k'l'} = 1.$$

Putting $\tau = i/\sqrt{3}$, we get

$$(34.82) \qquad\qquad k'(i/\sqrt{3}) = \frac{1}{2}\sqrt{2 - \sqrt{3}}.$$

35. A set of functional equations. We shall now study the function

$$(35.1) \qquad G(b,t,q) \equiv \prod_{n\geq 1} \frac{(1-q^n)^2(1-btq^{n-1})(1-b^{-1}t^{-1}q^n)}{(1-bq^{n-1})(1-b^{-1}q^n)(1-tq^{n-1})(1-t^{-1}q^n)}.$$

By (18.6), we may write

$$(35.11) \qquad G(b,t,q) = \frac{1-bt}{(1-b)(1-t)} + \sum_{n,k\geq 1} q^{kn}(b^k t^n - b^{-k}t^{-n}).$$

We observe that $G(t,b,q) = G(b,t,q)$, $G(b^{-1},t^{-1},q) = -G(b,t,q)$.

Now consider the odd part of $G(b,t,q)$, as a function of q. From (35.11), this is

$$\frac{1}{2}(G(b,t,q) - G(b,t,-q)) = \sum_{k,n \text{ odd}} q^{kn}(b^k t^n - b^{-k}t^{-n})$$

$$= \sum_{k,n\geq 0} q^{(2k+1)(2n+1)}b^{2k+1}t^{2n+1} - \sum_{k,n\geq 1} q^{(2k-1)(2n-1)}b^{-2k+1}t^{-2n+1}$$

$$= btq\left\{1 + \sum_{k\geq 1} q^{2k}b^{2k} + \sum_{n\geq 1} q^{2n}t^{2n}\right.$$

$$\left. + \sum_{k,n\geq 1} q^{4kn}((b^2q^2)^k(t^2q^2)^n - (b^2q^2)^{-k}(t^2q^2)^{-n})\right\}$$

$$= btq\left\{\frac{1-b^2t^2q^4}{(1-b^2q^2)(1-t^2q^2)} + \sum_{k,n\geq 1} q^{4kn}((b^2q^2)^k(t^2q^2)^n - (b^2q^2)^{-k}(t^2q^2)^{-n})\right\}.$$

Comparing with (35.11), we see that

$$(35.12) \qquad G(b,t,q) - G(b,t,-q) = 2btqG(b^2q^2, t^2q^2, q^4).$$

It follows immediately that

(35.13)
$$G(b,t,q) - G(b,t,-q) = G(-b,t,-q) - G(-b,t,q)$$
$$= G(b,-t,-q) - G(b,-t,q) = G(-b,-t,q) - G(-b,-t,-q).$$

Now take the even part of $G(b,t,q)$, as a function of b:

$$\frac{1}{2}(G(b,t,q) + G(-b,t,q))$$

$$= \frac{1 - b^2 t}{(1 - b^2)(1 - t)} + \sum_{k,n \geq 1} q^{2kn}(b^{2k}t^n - b^{-2k}t^{-n}),$$

that is,

(35.2)
$$G(b,t,q) + G(-b,t,q) = 2G(b^2,t,q^2).$$

For $t = -1$ we get the important special case

(35.21)
$$G(b,-1,q) + G(-b,-1,q) = 2G(b^2,-1,q^2),$$

which is the same as (31.14).

Next, we consider

$$\frac{1}{2}(G(b,t,q) - G(-b,t,-q))$$

$$= \frac{b}{1 - b^2} + \sum_{k,n \geq 1} \left(\frac{1 - (-1)^{k(n+1)}}{2}\right) q^{kn}(b^k t^n - b^{-k}t^{-n})$$

$$= \frac{b}{1 - b^2} + \sum_{\substack{k \text{ odd} \\ n \text{ even}}} q^{kn}(b^k t^n - b^{-k}t^{-n})$$

$$= \frac{b}{1 - b^2} + \sum_{\substack{k \geq 0 \\ n \geq 1}} q^{(2k+1)2n}b^{2k+1}t^{2n} - \sum_{k,n \geq 1} q^{(2k-1)2n}b^{-2k+1}t^{-2n}$$

$$= \frac{b}{1 - b^2} + b\sum_{n \geq 1} q^{2n}t^{2n} + b\sum_{k,n \geq 1} q^{4kn}(b^{2k}q^{2n}t^{2n} - b^{-2k}q^{-2n}t^{-2n}).$$

Thus we have proved

(35.3)
$$G(b,t,q) - G(-b,t,-q) = 2bG(b^2, t^2q^2, q^4),$$

with the special case

(35.31)
$$G(b,-1,q) - G(-b,-1,-q) = 2bG(b^2, q^2, q^4).$$

36. Application of (35.13). From now on, let p denote an odd integer, $p = 2\nu + 1$, $\nu > 0$. For $m = 1, 2, \ldots, \nu$, we define the functions

(36.1)
$$A_m \equiv A_m(q) \equiv A_m(q;p) = \prod_{n \geq 1} \frac{(1 - q^{pn})^2}{(1 - q^{pn-2m})(1 - q^{pn-(p-2m)})},$$

$$B_m \equiv B_m(q) \equiv B_m(q;p) = \prod_{n \geq 1} \frac{(1 - q^{pn})^2}{(1 + q^{pn-2m})(1 + q^{pn-(p-2m)})},$$

$$C_m(q) = A_m(-q); \qquad D_m(q) = B_m(-q).$$

Furthermore, for *arbitrary* m, let

(36.11)
$$P_m \equiv P_m(q) \equiv P_m(q;p) = \prod_{n \geq 1} \frac{(1-q^{pn})^2}{(1-q^{pn-2m})(1-q^{pn-(p-2m)})},$$

$$Q_m \equiv Q_m(q) \equiv Q_m(q;p) = \prod_{m \geq 1} \frac{(1-q^{pn})^2}{(1+q^{pn-2m})(1+q^{pn-(p-2m)})}.$$

Thus, for $1 \leq m \leq \nu$, $P_m = A_m$, $Q_m = B_m$. For $\nu < m < p$, however, we write $j = p - m$, so that $1 \leq j \leq \nu$, and elementary considerations yield

(36.12)
$$P_m(q) = -q^{p-2j}A_j(q); \qquad P_m(-q) = q^{p-2j}C_j(q),$$
$$Q_m(q) = q^{p-2j}B_j(q); \qquad Q_m(-q) = -q^{p-2j}D_j(q).$$

Now we return to $G(b,t,q)$ and set $b = \pm q^{2M}$, $t = \pm q^{2N}$, $1 \leq M, N \leq \nu$, after replacing q by $\pm q^p$. An examination of the eight cases leads to the following table:

(36.2)
$$G(q^{2M}, q^{2N}, q^p) = \frac{A_M A_N}{P_{M+N}(q)}; \qquad G(q^{2M}, q^{2N}, -q^p) = \frac{C_M C_N}{P_{M+N}(-q)},$$
$$G(-q^{2M}, q^{2N}, q^p) = \frac{B_M A_N}{Q_{M+N}(q)}; \qquad G(-q^{2M}, q^{2N}, -q^p) = \frac{D_M C_N}{Q_{M+N}(-q)},$$
$$G(q^{2M}, -q^{2N}, q^p) = \frac{A_M B_N}{Q_{M+N}(q)}; \qquad G(q^{2M}, -q^{2N}, -q^p) = \frac{C_M D_N}{Q_{M+N}(-q)},$$
$$G(-q^{2M}, -q^{2N}, q^p) = \frac{B_M B_N}{P_{M+N}(q)}; \qquad G(-q^{2M}, -q^{2N}, -q^p) = \frac{D_M D_N}{P_{M+N}(-q)}.$$

To reduce the P and Q factors, we distinguish two cases:

(36.21)
$$M + N \leq \nu; \quad \varepsilon = +1, \quad j = M + N,$$
$$M + N > \nu; \quad \varepsilon = -1, \quad j = p - (M + N).$$

Now replace b by q^{2M}, t by q^{2N}, after $q \to q^p$, in (35.13). After simplification by means of (36.21), (36.2), and (36.12), we obtain the set of equations

(36.3)
$$\frac{A_M A_N}{A_j} - \varepsilon \frac{C_M C_N}{C_j} = \frac{D_M C_N}{D_j} - \varepsilon \frac{B_M A_N}{B_j}$$
$$= \frac{C_M D_N}{D_j} - \varepsilon \frac{A_M B_N}{B_j} = \frac{B_M B_N}{A_j} - \varepsilon \frac{D_M D_N}{C_j}.$$

Here we introduce the functions

(36.4)
$$x_M \equiv x_M(\tau) = \prod_{n \geq 1} \frac{(1-q^{pn-2M})(1-q^{pn-(p-2M)})}{(1+q^{pn-2M})(1+q^{pn-(p-2M)})},$$

$$y_M \equiv y_M(\tau) = x_M(\tau + 1) \qquad (1 \leq M \leq \nu).$$

Clearly $x_M = B_M/A_M$, $y_M = D_M/C_M$. Further, let $u_M = \sqrt{x_M}$, $v_M = \sqrt{y_M}$, the sign being chosen positive for τ purely imaginary.

We propose to show that all the ratios B_M/A_M, B_M/C_M, and so on, can be expressed simply in terms of u_M, v_M, and $\lambda = \vartheta_4(p\tau)/\vartheta_3(p\tau)$, $\lambda^2 = l'$.

Separating the factors in the products for B_M, C_M, according as the exponents are odd or even, we find

(36.41) $$B_M/C_M = \vartheta_4^2(p\tau)\Phi_M(2\tau),$$

where

(36.42) $$\Phi_M(\tau) = \prod_{n\geq 1} \frac{(1+q^{pn})^2(1-q^{pn-M})(1-q^{pn-(p-M)})}{(1-q^{pn})^2(1+q^{pn-M})(1+q^{pn-(p-M)})}.$$

Denote by M' that one of $M/2$, $(p-M)/2$ which is an integer. The correspondence between M and M' is one-to-one; in fact, $M = \min(2M', p - 2M')$. By definition of x_M, (36.42) may be written

(36.43) $$\Phi_M(\tau) = \frac{x_{M'}(\tau)}{\vartheta_4^2(p\tau)},$$

so that (36.41) becomes

(36.44) $$\frac{B_M}{C_M} = \frac{\theta_4^2(p\tau)}{\vartheta_4^2(2p\tau)}x_{M'}(2\tau).$$

Replacing q by $-q$, that is, τ by $\tau + 1$, we have

(36.45) $$\frac{D_M}{A_M} = \frac{\vartheta_3^2(p\tau)}{\vartheta_4^2(2p\tau)}x_{M'}(2\tau),$$

and forming the ratio,

(36.46) $$\frac{A_M B_M}{C_M D_M} = \frac{\vartheta_4^2(p\tau)}{\vartheta_3^2(p\tau)} = l' = \lambda^2.$$

The elimination of $B_M = A_M x_M$, $D_M = C_M y_M$ yields

$$\lambda^2 = \frac{A_M^2}{C_M^2}\cdot\frac{x_M}{y_M}, \qquad \frac{A_M}{C_M} = \lambda\sqrt{\frac{y_M}{x_M}} = \lambda\frac{v_M}{u_M}.$$

It is now easy to construct the following table:

(36.47)
$$\frac{B_M}{A_M} = u_M^2, \quad \frac{B_M}{C_M} = \lambda u_M v_M, \quad \frac{B_M}{D_M} = \lambda\frac{u_M}{v_M}, \quad \frac{A_M}{B_M} = \frac{1}{u_M^2},$$
$$\frac{C_M}{A_M} = \frac{1}{\lambda}\frac{u_M}{v_M}, \quad \frac{A_M}{C_M} = \lambda\frac{v_M}{u_M}, \quad \frac{C_M}{D_M} = \frac{1}{v_M^2}, \quad \frac{C_M}{B_M} = \frac{1}{\lambda u_M v_M},$$
$$\frac{D_M}{A_M} = \frac{1}{\lambda}u_M v_M, \quad \frac{D_M}{C_M} = v_m^2, \quad \frac{A_M}{D_M} = \frac{\lambda}{u_M v_M}, \quad \frac{D_M}{B_M} = \frac{1}{\lambda}\frac{v_M}{u_M}.$$

By means of (36.47), the equations (36.3) may be reduced to the following form:

(36.51) $$\frac{v_j}{u_j} = \varepsilon\lambda\frac{(u_M u_N - u_M^{-1} u_N^{-1})}{(v_M v_N - v_M^{-1} v_N^{-1})} \qquad (1 \leq M, N \leq \nu),$$

(36.52) $$\frac{v_j}{u_j} = \frac{\varepsilon}{\lambda}\frac{(v_M v_N^{-1} - v_N v_M^{-1})}{(u_M u_N^{-1} - u_N u_M^{-1})} \qquad (M \neq N),$$

(36.53) $$\frac{v_N}{u_N} = \frac{\varepsilon}{\lambda}\frac{(v_M v_j^{-1} + \varepsilon v_j v_M^{-1})}{(u_M u_j^{-1} + \varepsilon u_j u_M^{-1})} \qquad (j \neq M).$$

From the product definitions, we have

$$(36.54) \qquad \prod_{M=1}^{\nu} u_M^2 = \prod_{M=1}^{\nu} x_M = \frac{\vartheta_4(\tau)}{\vartheta_4(p\tau)} = \sqrt{\frac{k'K}{l'L}} = \frac{\kappa}{\lambda}\rho,$$

where $\kappa = \sqrt{k'}$, $\rho = \sqrt{R}$, with the usual choice of the square roots. Similarly,

$$(36.55) \qquad \prod_{M=1}^{\nu} v_M^2 = \prod_{M=1}^{\nu} y_M = \frac{\vartheta_3(\tau)}{\vartheta_3(p\tau)} = \rho.$$

37. Two modular equations. We pause to show how (36.51)–(36.55) lead to interesting forms of the modular equation in the cases $p = 3$ and $p = 5$.

For $p = 3$, we put $M = N = 1$ in (36.51), $\varepsilon = -1$. Then

$$\frac{v_1}{u_1} = -\lambda \frac{(u_1^2 - u_1^{-2})}{(v_1^2 - v_1^{-2})} = -\lambda \frac{v_1^2 (1 - u_1^4)}{u_1^2 (1 - v_1^4)}.$$

In view of (36.54) and (36.55), $u_1^2 = \kappa\rho/\lambda$, $v_1^2 = \rho$, so

$$(\kappa/\lambda)^{1/2} = -\lambda \frac{(1 - k'R/l')}{(1 - R)},$$

and solving for R, we get

$$(37.1) \qquad R = \frac{l' + (k'l')^{1/4}}{k' + (k'l')^{1/4}};$$

with $\tau \to -1/3\tau$,

$$(37.2) \qquad \frac{3}{R} = \frac{k + (kl)^{1/4}}{l + (kl)^{1/4}}.$$

Turning to the case $p = 5$, we select $M = N = 1$, $j = 2$, $\varepsilon = +1$; $M = 1$, $N = 2$, $j = 2$, $\varepsilon = -1$ in (36.51), and $M = N = 1$, $j = 2$, $\varepsilon = +1$ in (36.53). These yield

$$\frac{v_2}{u_2} = \lambda \frac{(u_1^2 - u_1^{-2})}{(v_1^2 - v_1^{-2})},$$

$$\frac{v_2}{u_2} = -\lambda \frac{(u_1 u_2 - u_1^{-1} u_2^{-1})}{(v_1 v_2 - v_1^{-1} v_2^{-1})},$$

$$\frac{v_1}{u_1} = \frac{1}{\lambda} \frac{(v_1 v_2^{-1} + v_2 v_1^{-1})}{(u_1 u_2^{-1} + u_2 u_1^{-1})}.$$

These equations, together with

$$u_1^2 u_2^2 = \frac{\kappa\rho}{\lambda}, \qquad v_1^2 v_2^2 = \rho,$$

enable us to solve easily for R:

$$(37.3) \qquad R = \frac{l' + (k'/l')^{1/4}}{k' + (k'/l')^{1/4}},$$

$$(37.4) \qquad \frac{5}{R} = \frac{k + (l/k)^{1/4}}{l + (l/k)^{1/4}}.$$

The similarity of (37.1) and (37.3) is noteworthy. Neither is particularly suitable for computation of the singular modulus $k'(i/\sqrt{p})$.

38. Continuation of §36. By using the reduction $\vartheta_4^2(2\tau) = \vartheta_3(\tau)\vartheta_4(\tau)$ ((34.7)), together with the formula for B_M/C_M in (36.47), we may reduce (36.44) to the form

$$(38.1) \qquad\qquad x_M y_M = t_{M'}$$

where we have set

$$(38.11) \qquad\qquad t_M(\tau) = x_M^2(2\tau).$$

Now we turn to (35.2), replace q by q^p, b by q^{2M}. The three functions involved are

$$(38.21) \qquad\qquad G(q^{2M}, -1, q^p) = \frac{1}{2}\vartheta_4^2(p\tau)\frac{P_M(q)}{Q_M(q)},$$

$$(38.22) \qquad\qquad G(-q^{2M}, -1, q^p) = \frac{1}{2}\vartheta_4^2(p\tau)\frac{Q_M(q)}{P_M(q)}$$

$$(38.23) \qquad\qquad G(q^{4M}, -1, q^{2p}) = g_M(q^2),$$

where

$$g_M(q) = \frac{1}{2}\vartheta_4^2(p\tau)\frac{P_M(q)}{Q_M(q)},$$

by (38.21). Thus (35.2) becomes

$$\vartheta_4^2(p\tau)\left(\frac{P_M(q)}{Q_M(q)} + \frac{Q_M(q)}{P_M(q)}\right) = 2\vartheta_4^2(2p\tau)\frac{P_M(q^2)}{Q_M(q^2)},$$

or, if $1 \le M \le \nu$,

$$\vartheta_4^2(p\tau)\left(\frac{A_M(q)}{B_M(q)} + \frac{B_M(q)}{A_M(q)}\right) = 2\vartheta_4^2(2p\tau)\frac{A_M(q^2)}{B_M(q^2)}.$$

Finally, using (34.7) again, together with the definitions of x_M and t_M, we obtain

$$(38.3) \qquad\qquad x_M + x_M^{-1} = 2/\lambda\sqrt{t_M},$$

and on replacing τ by $\tau + 1$, so that $x_M \to y_M$, $\lambda \to \lambda^{-1}$, $t_M \to t_M$,

$$(38.31) \qquad\qquad y_M + y_M^{-1} = 2\lambda/\sqrt{t_M}.$$

Now replace x_M^{-1} in (38.3) and y_M^{-1} in (38.31) by their respective values from (38.1) to obtain the two equations

$$(38.32) \qquad\qquad x_M + t_{M'}^{-1}y_M = 2/\lambda\sqrt{t_M},$$
$$(38.33) \qquad\qquad t_{M'}^{-1}x_M + y_M = 2\lambda/\sqrt{t_M}.$$

Solving these as linear equations in x_M, y_M, we obtain

$$(38.34) \qquad\qquad (1 - t_{M'}^{-2})x_M = \frac{2}{\lambda\sqrt{t_M}}\left(1 - \frac{\lambda^2}{t_{M'}}\right),$$

$$(38.35) \qquad\qquad (1 - t_{M'}^{-2})y_M = \frac{2}{\lambda\sqrt{t_M}}\left(\lambda^2 - \frac{1}{t_{M'}}\right).$$

Multiplying these two equations and using (38.1) again, then solving for t_M, we get

$$(38.4) \qquad t_M = \frac{4t_{M'}(1 - \lambda^{-2}t_{M'})(1 - \lambda^2 t_{M'})}{(1 - t_{M'}^2)^2} \equiv g(t_{M'}).$$

Equation (38.4) is of considerable interest. It shows that the functions t_M, $1 \le M \le \nu$, are rationally related in certain cycles, over the base field Γ of the rationals with $\gamma = l' + 1/l'$ adjoined. The cycles are determined by the substitution $T: M' \to M = \min(2M', p - 2M')$. The powers of T form a subgroup of the symmetric group on ν letters, and the order ω of the subgroup is given by the least positive integer satisfying $2^\omega \equiv \pm 1 \pmod{p}$. If we write g_m for the mth iterate of the rational function g, we have $t_M = g_\omega(t_M)$. Thus the functions t_M are algebraic over the field Γ.

If we define

$$(38.41) \qquad H(z) = \prod_{M=1}^{\nu} (1 - z t_M) = 1 - \sigma_1 z + \sigma_2 z^2 - \cdots + (-1)^\nu \sigma_\nu z^\nu,$$

the product σ_ν is of particular interest to us, for by (38.1), (36.54), and (36.55),

$$(38.42) \qquad \sigma_\nu = \prod_{M'=1}^{\nu} t_{M'} = \prod_{M=1}^{\nu} x_M y_M = \frac{\kappa}{\lambda} \rho^2 = \left(\frac{k'}{l'}\right)^{1/2} R.$$

Thus, to find the modular equation of degree p, it is sufficient to determine ν values of $H(z)$, or relations among the values sufficient to determine the coefficients σ_i.

As an example, divide (38.35) by (38.34):

$$(38.5) \qquad \frac{y_M}{x_M} = \frac{1}{l'}\left(\frac{1 - l't_{M'}}{1 - t_{M'}/l'}\right).$$

Multiplying for $M = 1, 2, \ldots, \nu$, hence for M' over the same range, and using (36.54) and (36.55), we find

$$(38.51) \qquad \frac{H(l')}{H(1/l')} = \frac{\lambda^p}{\kappa} = (l'^p/k')^{1/2} \equiv \mu'.$$

This is already enough for $p = 3$, where

$$H(l') = 1 - l'\sigma_1, \qquad H(1/l') = 1 - \sigma_1/l'.$$

Solving for $\sigma_1 = R\sqrt{k'/l'}$, we obtain

$$(38.52) \qquad R = \frac{l' - \sqrt{k'/l'}}{\sqrt{k'/l'} - k'}, \qquad \frac{3}{R} = \frac{k - \sqrt{l/k}}{\sqrt{l/k} - l}.$$

We can actually determine the individual values $H(l')$, $H(1/l')$, however. For this purpose, we refer to (35.31), making the customary substitutions to obtain

$$(38.53) \qquad G(q^{2M}, -1, q^p) - G(-q^{2M}, -1, q^p) = 2q^{2M} G(q^{4M}, q^{2p}, q^{4p}).$$

If we multiply over $M = 1, 2, \ldots, \nu$, the product on the right reduces to

$$(38.54) \qquad 2^{-\nu} \vartheta_2^p(p\tau)/\vartheta_2(\tau);$$

by (38.21) and (38.22), together with the table (36.47), the product on the left becomes

$$(38.55) \qquad 2^{-\nu} \prod_{M=1}^{\nu} \left\{ \frac{\vartheta_4^2(p\tau)}{x_M} - \vartheta_3^2(p\tau) y_M \right\}$$

$$= 2^{-\nu} \vartheta_4^{2\nu}(p\tau) \left(\prod_{M=1}^{\nu} x_M \right)^{-1} \prod_{M=1}^{\nu} \left(1 - \frac{1}{l'} x_M y_M \right)$$

$$= 2^{-\nu} \vartheta_4^{2\nu}(p\tau) \frac{\lambda}{\kappa \rho} H(1/l').$$

We recall that

$$(38.56) \qquad \begin{aligned} \vartheta_2^2(\tau) &= \frac{2kK}{\pi}, & \vartheta_2^2(p\tau) &= \frac{2lL}{\pi}, \\ \vartheta_4^2(\tau) &= \frac{2k'K}{\pi}, & \vartheta_4^2(p\tau) &= \frac{2l'L}{\pi}. \end{aligned}$$

Using these, and equating (38.54) and (38.55), we obtain

$$(38.57) \qquad H(1/l') = \frac{(l^p/k)^{1/2}}{(l'^p/k')^{1/2}}.$$

This, together with (39.51), yields

$$(38.58) \qquad H(l') = (l^p/k)^{1/2} \equiv \mu.$$

The value (38.58) could also have been derived directly from (38.57) by the transformation $\tau \to \tau + 1$, under which t_M is invariant, and

$$(38.59) \qquad \begin{aligned} k' &\to \frac{1}{k'}, & l' &\to \frac{1}{l'}, \\ k &\to \frac{ik}{k'}, & l &\to i^p \frac{l}{l'} = (-1)^{\nu} \left(\frac{il}{l'} \right), & l^p &\to \frac{il^p}{l'^p}. \end{aligned}$$

The use of (38.57) and (38.58) to determine R in the case $p = 3$ leads us to

$$(38.6) \qquad \begin{aligned} R &= (l'^3/k')^{1/2} - (l^3/k)^{1/2}, \\ R &= (k'l')^{-1/2} (1 - (l^3/k)^{1/2}), \end{aligned}$$

and equating the two forms in (38.6) again gives us $(kl)^{1/2} + (k'l')^{1/2} = 1$, which we obtained previously.

For $p = 5$ we have the two equations

$$1 - \sigma_1 l' + \sigma_2 l'^2 = \mu,$$
$$1 - \sigma_1/l' + \sigma_2/l'^2 = \mu/\mu'.$$

Solving for σ_2, then R, we get

$$(38.61) \qquad \begin{aligned} R &= (l/k)^{1/2} + (l'/k')^{1/2} - (ll'/kk')^{1/2}, \\ \frac{5}{R} &= (k/l)^{1/2} - (k'/l') - (kk'/ll')^{1/2}. \end{aligned}$$

With $u = (l/k)^{1/2}$, $v = (l'/k')^{1/2}$, elimination of R yields

$$(38.62) \qquad (u - v)^2 = (u + v)(1 + uv).$$

39. Other functional values of $H(z)$. By multiplying all the equations (38.3) together, we get

$$\prod_{M=1}^{\nu} (1 + x_M^2(\tau)) = \left(\frac{2}{\lambda}\right)^{\nu} \left(\prod x_M\right) \left(\prod t_M\right)^{-1/2}$$

(39.1)
$$= \left(\frac{2}{\lambda}\right)^{\nu} \left(\frac{\kappa\rho}{\lambda}\right) \left(\frac{\kappa\rho^2}{\lambda}\right)^{-1/2}$$

$$= \left(\frac{2}{\lambda}\right)^{\nu} \left(\frac{\kappa}{\lambda}\right)^{1/2} \equiv a(\tau).$$

Therefore

(39.11)
$$a(2\tau) = \prod_{M=1}^{\nu} (1 + x_M^2(2\tau)) = \prod_{M=1}^{\nu} (1 + t_M(\tau)) = H(-1).$$

Now the modular equation of degree 2, (34.76), yields

(39.12)
$$k'(2\tau) = \frac{2\sqrt{k'}}{1 + k'}, \qquad l'(2\tau) = \frac{2\sqrt{l'}}{1 + l'},$$

so that

$$a(2\tau) = 2^{\nu} \left(\frac{k'(2\tau)}{l'^p(2\tau)}\right)^{1/4} = \frac{2^{\nu/2}}{\mu'^{1/4}} \left(\frac{(1 + l')^p}{1 + k'}\right)^{1/4}.$$

Thus we have obtained a third value

(39.13)
$$H(-1) = \left\{ \frac{2^{p-1}}{\mu'} \left(\frac{(1 + l')^p}{(1 + k')}\right) \right\}^{1/4}.$$

Multiplying all the equations (38.4) gives us

(39.14)
$$(H(1)H(-1))^2 = 2^{p-1} H(l')H(1/l') = 2^{p-1} \mu^2/\mu'.$$

The combination of (39.14) and (39.13), and simplification by means of $k^2 + k'^2 = l^2 + l'^2 = 1$, produces

(39.15)
$$H^2(1) = \left\{ \frac{2^{p-1}}{\mu'} \left(\frac{(1 - l')^p}{(1 - k')}\right) \right\}^{1/2}.$$

To determine the proper sign for the square root, we write down the leading terms in the expansions of the various functions:

$$x_M = 1 - 2q^{\rho_M} + \cdots, \qquad \rho_M = \min(2M, p - 2M),$$

$$t_M = 1 - 4q^{2\rho_M} + \cdots,$$

$$k'^{1/2} = \frac{\vartheta_4(\tau)}{\vartheta_3(\tau)} = \prod_1^\infty \frac{(1 - q^{2n-1})^2}{(1 + q^{2n-1})^2} = 1 - 4q + \cdots,$$

$$k'^{1/4} = \prod_1^\infty \frac{(1 - q^{2n-1})}{(1 + q^{2n-1})} = 1 - 2q + \cdots,$$

$$(39.16) \qquad k^{1/2} = \frac{\vartheta_2(\tau)}{\vartheta_3(\tau)} = 2q^{1/4} \prod_1^\infty \frac{(1 - q^n)^2 (1 - q^{4n})^4}{(1 - q^{2n})^6} = 2q^{1/4}(1 - 2q + \cdots),$$

$$k^{1/4} = \sqrt{2} q^{1/8}(1 - q + \cdots),$$

$$l'^{1/4} = 1 - 2q^p + \cdots,$$

$$l^{1/4} = \sqrt{2} q^{p/8}(1 - q^p + \cdots),$$

$$\mu' = (l'^{1/2})^p / k'^{1/2} = 1 + 4q + \cdots,$$

$$\mu = (l^{1/2})^p / k^{1/2} = 2^{p-1} q^{(p^2 - 1)/4} + \cdots,$$

$$\mu^{1/4} = 2^{(p-1)/4} q^{(p^2 - 1)/16} + \cdots.$$

From (39.16), we have

$$H(1) = \prod_{M=1}^\nu (1 - t_M) = \prod_{M=1}^\nu (4q^{2\rho_M} + \cdots) = 2^{p-1} q^{2\Sigma\rho_M} + \cdots,$$

so that

$$(39.17) \qquad H(1) = \left\{ \frac{2^{p-1}}{\mu'} \left(\frac{(1 - l')^p}{(1 - k')} \right) \right\}^{1/4}.$$

We shall now set forth a device by means of which it is possible to obtain new relations for H from known functional values. Suppose that we know

$$(39.2) \qquad H(z(\tau)|\tau) \equiv \prod_{M=1}^\nu (1 - z(\tau) t_M(\tau)) \equiv \phi(\tau).$$

Write $\varsigma(\tau) = z(\tau/2)$, $\psi(\tau) = \phi(\tau/2)$, and replace τ by $\tau/2$ in (39.2):

$$(39.21) \qquad \psi(\tau) = \prod_{M=1}^\nu (1 - \varsigma(\tau) x_M^2(\tau)),$$

since $t_M(\tau) = x_M^2(2\tau)$. From (38.1) and (38.5),

$$(39.22) \qquad x_M^2 = \frac{t_{M'}(l' - t_{M'})}{(1 - l' t_{M'})},$$

so that, after the primes are dropped, (39.21) becomes

$$(39.23) \qquad \begin{aligned} \psi(\tau) &= \prod_{M=1}^\nu \left\{ \frac{1 - l'(1 + \varsigma) t_M + t_M^2}{1 - l' t_M} \right\} \\ &= \frac{H(\alpha_1) H(\alpha_2)}{H(l')} = \frac{H(\alpha_1) H(\alpha_2)}{\mu}, \end{aligned}$$

where α_1, α_2 are determined by

$$\alpha_1 + \alpha_2 = l'(1 + \varsigma), \qquad \alpha_1 \alpha_2 = \varsigma,$$

that is,

(39.24)
$$\alpha_1, \alpha_2 = \tfrac{1}{2}\{l'(1+\varsigma) \pm \sqrt{\Delta}\},$$
$$\Delta = l'^2(1+\varsigma)^2 - 4\varsigma.$$

The necessary transformation formulas are

(39.25)
$$k'(\tau/2) = \frac{1-k}{1+k}, \qquad k(\tau/2) = \frac{2\sqrt{k}}{1+k},$$
$$l'(\tau/2) = \frac{1-l}{1+l}, \qquad l(\tau/2) = \frac{2\sqrt{l}}{1+l}.$$

As a first example, take $z = \varsigma = 1$, so that $\alpha_{1,2} = l' \pm il$. By (39.17),

$$\phi(\tau) = H(1) = \left\{ \frac{2^{p-1}}{\mu'} \frac{(1-l')^p}{(1-k')} \right\}^{1/4},$$
$$\psi(\tau) = 2^{(p-1)/2}(\mu/\mu')^{1/2},$$

and therefore

(39.26) $H(l' + il)H(l' - il) = 2^{(p-1)/2}(\mu^{3/2}/\mu'^{1/2}).$

Next, take $z = l'$, $\varsigma = (1-l)/(1+l)$, so that $\alpha_1 = \alpha_2 = (1-l)/l'$. By (38.58),

$$\phi(\tau) = H(l') = \mu,$$

(39.27)
$$\psi(\tau) = \mu(\tau/2) = \left\{ 2^{p-1}\mu \frac{(1+k)}{(1+l)^p} \right\}^{1/2},$$
$$H^2\left(\frac{1-l}{l'}\right) = \left\{ 2^{p-1}\mu^3 \frac{(1+k)}{(1+l)^p} \right\}^{1/2}.$$

Now, by (39.16),

$$\frac{1-l}{l'} = 1 - 4q^{p/2} + \cdots,$$

$$H\left(\frac{1-l}{l'}\right) = \prod_{M=1}^{\nu} \{1 - (1 - 4q^{p/2} + \cdots)(1 - 4q^{2\rho M} + \cdots)\}$$
$$= 2^{p-1}q^{\Sigma c_M} + \cdots,$$

where $c_M = \min(2\rho_M, p/2) = \min(4M, 2p - 4M, p/2)$. Hence

(39.28) $H\left(\dfrac{1-l}{l'}\right) = \left\{ 2^{p-1}\mu^3 \dfrac{(1+k)}{(1+l)^p} \right\}^{1/4},$

the leading term being

(39.29) $2^{p-1}q^{\Sigma c_M} = 2^{p-1}q^{3(p^2-1)/16},$

since $\mu = 2^{p-1}q^{(p^2-1)/4} + \cdots, 1 + k = 1 + \cdots, 1 + l = 1 + \cdots.$

Now by (38.59), under $\tau \to \tau + 1 \to \tau + 2 \to \tau + 3$, we have

(39.3)

$$\frac{1-l}{l'} \to l' - i^p l \to \frac{1+l}{l'} \to l' + i^p l,$$

$$\mu^{1/4} \to i^{(p^2-1)/8} \left(\frac{\mu}{\mu'}\right)^{1/4} \to (-1)^{(p^2-1)/8} \mu^{1/4} \to (-i)^{(p^2-1)/8} \left(\frac{\mu}{\mu'}\right)^{1/4},$$

$$\frac{(1+k)}{(1+l)^p} \to \frac{(k'+ik)\mu'^2}{(l'+i^p l)^p} \to \frac{(1-k)}{(1-l)^p} \to \frac{(k'-ik)\mu'^2}{(l'-i^p l)^p}.$$

Since $t_M(\tau + 1) = t_M(\tau)$, this yields three new values for H:

(39.31)
$$H(l' - i^p l) = (-i)^{(p^2-1)/8} \left\{ 2^{p-1} \frac{\mu^3}{\mu'} \frac{(k'+ik)}{(l'+i^p l)^p} \right\}^{1/4},$$

(39.32)
$$H\left(\frac{1+l}{l'}\right) = (-1)^{(p^2-1)/8} \left\{ 2^{p-1} \mu^3 \frac{(1-k)}{(1-l)^p} \right\}^{1/4},$$

(39.33)
$$H(l' + i^p l) = i^{(p^2-1)/8} \left\{ 2^{p-1} \frac{\mu^3}{\mu'} \frac{(k'-ik)}{(l'-i^p l)^p} \right\}^{1/4}.$$

As a check, the product of (39.31) and (39.33) recovers (39.26).

The eight values of H computed so far would enable us to find the modular equations of degrees up to 17 by linear methods. Those of lower degree could appear in many different forms. It would be of interest to find a general method for computing functional values.

An alternate approach, to modular equations of a different kind, has been treated in [17].

40. A system of identities. We turn now to the development of a system of relations which exist among the functions $x_M(\tau)$, $M = 1, 2, \ldots, \nu$. For this purpose, we make the usual substitutions $q \to q^p$, $b \to q^{2M}$, $t \to q^{2N}$, $1 \leq M, N \leq \nu$, in (35.2), to obtain

(40.1)
$$G(q^{2M}, q^{2N}, q^p) + G(-q^{2M}, q^{2N}, q^p) = 2G(q^{4M}, q^{2N}, q^{2p}).$$

Reference to the table (36.2) and examination of the product on the right yield

(40.11)
$$\frac{A_M(q) A_N(q)}{P_{M+N}(q)} + \frac{B_M(q) A_N(q)}{Q_{M+N}(q)} = 2 \frac{P_M(q^2) P_{N'}(q^2)}{P_{(2M+N)'}(q^2)},$$

where a', as usual, denotes that one of $a/2$, $(p-a)/2$ which is an integer. Now, for any m,

$$P_m(q^2) = \prod_{n \geq 1} \frac{(1 - q^{2pn})^2}{(1 - q^{2pn-4m})(1 - q^{2pn-2p+4m})}$$

(40.12)
$$= \prod \left(\frac{1 + q^{pn}}{1 - q^{pn}}\right)^2 \prod \frac{(1 - q^{pn})^2}{(1 - q^{pn-2m})(1 - q^{pn-p+2m})}$$

$$\cdot \prod \frac{(1 - q^{pn})^2}{(1 + q^{pn-2m})(1 + q^{pn-p+2m})}$$

$$= \vartheta_4^{-2}(p\tau) P_m(q) Q_m(q).$$

Thus (40.11) may be written

$$(40.13) \qquad A_N \left(\frac{B_M}{Q_{M+N}} + \frac{A_M}{P_{M+N}} \right) = 2c \frac{A_M B_M A_{N'} B_{N'}}{P_{(2M+N)'} Q_{(2M+N)'}},$$

where we have set $c = \vartheta_4^{-2}(p\tau)$.

Further reduction of (40.13) falls into cases. Writing $d = (2M + N)'$, we see that $d = N' + M$ if $N = 2N'$, and $d = N' - M$ if $N = p - 2N'$. In all cases $P_d Q_d$ reduces to $A_s B_s$ multiplied by an elementary factor, where $s = \min(|d|, p - |d|)$. The reductions are effected by means of

$$(40.14) \qquad \begin{aligned} P_m(q) &= -q^{-2m} P_{-m}(q), & Q_m(q) &= q^{-2m} Q_{-m}(q). \\ P_m(q) &= -q^{-p+2m} P_{p-m}(q), & Q_m(q) &= q^{-p+2m} Q_{p-m}(q), \end{aligned}$$

which are easily verified. Similarly, Q_{M+N} and P_{M+N} on the left of (40.13) reduce to B_j and A_j, where $j = \min(M + N, p - M - N)$; as in §36, we set $\varepsilon = +1, -1$, according as $M + N \leq \nu$ or $> \nu$. We may now distinguish six nontrivial cases, which we proceed to discuss:

Case I. $N = 2N'$, $M + N \leq \nu$.
Here $0 < d \leq \nu$, so $s = d = N' + M$. Also $\varepsilon = 1$, $j = M + N$. Thus (40.13) becomes

$$(40.21) \qquad A_N \left(\frac{B_M}{B_j} + \varepsilon \frac{A_M}{A_j} \right) = 2c \frac{A_M B_M A_{N'} B_{N'}}{A_s B_s}.$$

Case II. $N = 2N'$, $N' + M \leq \nu < N + M$.
Here $s = d = N' + M$, $\varepsilon = -1$, $j = p - (M + N)$, so

$$(40.22) \qquad A_N \left(\frac{B_M}{B_j} + \varepsilon \frac{A_M}{A_j} \right) = q^{p-2j} \cdot 2c \frac{A_M B_M A_{N'} B_{N'}}{A_s B_s}.$$

Case III. $N = 2N'$, $N' + M > \nu$.
Here $d > \nu$, so $s = p - d = p - (N' + M)$, $\varepsilon = -1$, $j = p - (M + N)$. Hence

$$(40.23) \qquad A_N \left(\frac{B_M}{B_j} + \varepsilon \frac{A_M}{A_j} \right) = -q^{p-2M} \cdot 2c \frac{A_M B_M A_{N'} B_{N'}}{A_s B_s}.$$

Case IV. $N = p - 2N'$, $M + N \leq \nu$.
Here $0 < N' - M = d \leq \nu$, so $s = d$. Also $\varepsilon = 1$, $j = M + N$, and

$$(40.24) \qquad A_N \left(\frac{B_M}{B_j} + \varepsilon \frac{A_M}{A_j} \right) = 2c \frac{A_M B_M A_{N'} B_{N'}}{A_s B_s}.$$

Case V. $N = p - 2N'$, $N' > M$, $M + N > \nu$.
Here, as in IV, $s = d = N' - M$, but $\varepsilon = -1$, $j = p - (M + N)$, so

$$(40.25) \qquad A_N \left(\frac{B_M}{B_j} + \varepsilon \frac{A_M}{A_j} \right) = q^{p-2j} \cdot 2c \frac{A_M B_M A_{N'} B_{N'}}{A_s B_s}.$$

Case VI. $N = p - 2N'$, $N' < M$.
Here $-\nu < N' - M = d < 0$, so $s = -d = M - N'$. Also, $M + N > N' + (p - 2N') = p - N' > \nu$, so $\varepsilon = -1$, $j = p - (M + N)$. Hence

$$(40.26) \qquad A_N \left(\frac{B_M}{B_j} + \varepsilon \frac{A_M}{A_j} \right) = -q^{p-2M} \cdot 2c \frac{A_M B_M A_{N'} B_{N'}}{A_s B_s}.$$

Recalling that $B_M = x_M A_M$, we may eliminate the B's in all these equations to obtain

(40.3) $$x_M + \varepsilon x_j = f_{M,N} \Phi_{M,N},$$

where

(40.31) $$\phi_{M,N} = 2c \frac{x_M x_j x_{N'}}{x_s} \cdot \frac{A_M A_j A_{N'}}{A_N A_s^2},$$

and s and $f = f_{M,N}$ are given by the following table:

$s = \min(M + N', p - (M + N'))$	$s = |N' - M|$

(40.32)

I. $\quad f = 1$	IV. $\quad f = 1$
II. $\quad f = q^{p-2j}$	V. $\quad f = q^{p-2j}$
III. $\quad f = -q^{p-2M}$	VI. $\quad f = -q^{p-2M}$

The case $d = 0$, which is not included, is trivial. For special values of p, some of the cases may also be trivial or redundant.

Let us now consider several examples. For $p = 3$, $M = N = 1$, equation (40.3) is trivial. For $p = 5$, $M = N = 1$ yields (Case IV)

(40.41) $$x_1 + x_2 = 2c x_2^2 A_1^{-2} A_2^3,$$

and $M = 1$, $N = 2$ (Case II):

(40.42) $$x_1 - x_2 = 2c q x_1^2 A_1^3 A_2^{-2}.$$

The pair $(M, N) = (2, 1)$ is trivial, and $(2,2)$ duplicates $(1,2)$. Dividing (40.42) by (40.41), we get

(40.43) $$\frac{x_1 - x_2}{x_1 + x_2} = q \left(\frac{x_1}{x_2} \right)^2 \left(\frac{A_1}{A_2} \right)^5.$$

Now let

(40.44) $$v(q) = \frac{A_1}{A_2} = \prod_{n=1}^{\infty} \frac{(1 - q^{5n-1})(1 - q^{5n-4})}{(1 - q^{5n-2})(1 - q^{5n-3})}.$$

We have

(40.45) $$\frac{x_2}{x_1} = \prod_{n=1}^{\infty} \frac{(1 - q^{5n-1})(1 - q^{5n-4})}{(1 + q^{5n-1})(1 + q^{5n-4})} \cdot \frac{(1 + q^{5n-2})(1 + q^{5n-3})}{(1 - q^{5n-2})(1 - q^{5n-3})}$$
$$= v(q) w(q),$$

where

(40.46) $$w(q) = \prod_{n \geq 1} \frac{(1 + q^{5n-2})(1 + q^{5n-3})}{(1 + q^{5n-1})(1 + q^{5n-4})}.$$

Now, $v(q)/w(q) = v(q^2)$, so $x_2/x_1 = v^2(q)/v(q^2)$, and (40.43) becomes

(40.47) $$\frac{v(q^2) - v^2(q)}{v(q^2) + v^2(q)} = q v(q) v^2(q^2).$$

Making the substitution $u(q) = q^{1/5}v(q)$ results in

(40.48)
$$\frac{u(q^2) - u^2(q)}{u(q^2) + u^2(q)} = u(q)u^2(q^2),$$

which is equivalent to a formula given by Mordell [31]. If we multiply the two equations (40.41), (40.42), we get

(40.49)
$$x_1^2 - x_2^2 = 4q \prod_{n \geq 1} \frac{(1 + q^{5n})^6(1 - q^n)}{(1 + q^n)^2(1 - q^{5n})} = 4\frac{\eta^2(\tau)\eta^6(10\tau)}{\eta^2(2\tau)\eta^7(5\tau)}.$$

For $p = 7$, the pairs $(M, N) = (1, 1)$, $(1, 2)$, $(2, 1)$, $(2, 2)$, and $(1, 3)$ yield the following equations, the four other pairs being trivial or redundant:

(40.51) $x_1 + x_2 = 2cx_1x_3 A_2^{-1} A_3^2$ (1, 1),

(40.52) $x_1 + x_3 = 2cx_1^2 x_2^{-1} x_3 A_1^3 A_2^{-3} A_3$ (1, 2),

(40.53) $x_2 + x_3 = 2cx_1^{-1} x_2 x_3^2 A_1^{-3} A_2 A_3^3$ (2, 1),

(40.54) $x_2 - x_3 = 2cqx_1 x_2 A_1^2 A_3^{-1}$ (2, 2),

(40.55) $x_1 - x_3 = 2cqx_2 x_3 A_1^{-1} A_2^2$ (1, 3).

Using these results, we can obtain the following modular relation, similar to (40.48):

(40.56) $R^{14}S^6(S+1)^2(S-1) - R^7 S^2(S-1)(2S^3 - 3S^2 - 2S - 1) + (S+1)^3 = 0$,

where

(40.57)
$$R(q) = q^{-1/7}\frac{A_1}{A_2} = q^{-1/7} \prod_{n \geq 1} \frac{(1 - q^{7n-3})(1 - q^{7n-4})}{(1 - q^{7n-2})(1 - q^{7n-5})},$$
$$S(q) = \frac{R(q^2)}{R^2(q)} = \frac{x_1}{x_2}.$$

It seems likely that relations of the form (40.48), (40.56) exist for every p.

By combining (40.51), (40.52), and (40.53), we get

$$\frac{(x_1 + x_3)(x_2 + x_3)}{(x_1 + x_2)^2} = \frac{x_3}{x_1},$$

which may be rewritten in the neat form

(40.58)
$$\frac{x_1}{x_3} - \frac{x_2}{x_1} + \frac{x_3}{x_2} = 1 \qquad (p = 7).$$

In order to systematize the discovery of identities like (40.58), we make correspond to every pair (M, N) for which (40.3) is nontrivial a vector $\varsigma = \varsigma(M, N)$:

(40.6) $\varsigma = (e_0, e_1, e_2, \ldots, e_\nu)$,

where the e_i $(i \geq 1)$ are defined by

(40.61) $\Phi_{M,N} = 2cx_M x_j x_{N'} x_s^{-1} A_1^{e_1} A_2^{e_2} \cdots A_\nu^{e_\nu}$,

from (40.31), and e_0 is given by

(40.62) $f_{M,N} = \pm q^{e_0}$.

Now suppose that we can find a linear relation

(40.63) $$\sum_r a_r \varsigma(M_r, N_r) = 0,$$

in which the a_r are integers not all zero. Then in the product

(40.64) $$\prod_r (x_{M_r} + \varepsilon_r x_{j_r})^{a_r} = \prod_r f_{M_r,N_r}^{a_r} \Phi_{M_r N_r}^{a_r},$$

from (40.3), neither side can involve an A factor or q. Furthermore, the exponent to which c appears is

(40.641) $$\sum_r a_r = \sum_r a_r \sum_{i=1}^{\nu} e_i^{(r)} = \sum_{i=1}^{\nu} \sum_r a_r e_i^{(r)} = 0,$$

where $e_i^{(r)}$ is the ith component of $\varsigma(M_r, N_r)$, and $\sum_{i=1}^{\nu} e_i^{(r)} = 1$, by (40.31). It follows that (40.64) can involve only the x_i, and that the dimension of each side in the variables x_i must by 0. Thus, if we write (40.64) in the form

(40.65) $$J(x_1, x_2, \ldots, x_\nu) = 0,$$

the function J, which we may assume to be a polynomial, is homogeneous in the x_i. Replacing τ by 2τ, and recalling that $x_i(2\tau) = \sqrt{t_i(\tau)}$, we have

(40.66) $$J(\sqrt{t_1}, \sqrt{t_2}, \ldots, \sqrt{t_\nu}) = 0.$$

But $\sqrt{t_i} = (2/\lambda)(x_i + x_i^{-1})^{-1}$, by (38.3); the homogeneity of J then implies

(40.67) $$J((x_1 + x_1^{-1})^{-1}, (x_2 + x_2^{-1})^{-1}, \ldots, (x_\nu + x_\nu^{-1})^{-1}) = 0.$$

For example, (40.58) goes over into

(40.68) $$\frac{x_3 + x_3^{-1}}{x_1 + x_1^{-1}} - \frac{x_1 + x_1^{-1}}{x_2 + x_2^{-1}} + \frac{x_2 + x_2^{-1}}{x_3 + x_3^{-1}} = 1 \qquad (p = 7).$$

In general, we shall write down only the primary relation (40.65).

For $p = 9$ we get two primary relations:

(40.7) $$\frac{x_1}{x_4} + \frac{x_2}{x_1} = \frac{x_3}{x_4} + \frac{x_3}{x_2} \qquad (p = 9),$$

(40.71) $$\frac{x_1}{x_4} + \frac{x_4}{x_3} + \frac{x_1}{x_3} = 1 + \frac{x_2}{x_3} + \frac{x_3}{x_2} \qquad (p = 9).$$

For $p = 11$ we get five relations:

(40.72) $$\frac{x_1}{x_4} + \frac{x_2}{x_1} = \frac{x_3}{x_4} + \frac{x_3}{x_2} \qquad (p = 11),$$

(40.73) $$\frac{x_1}{x_5} + \frac{x_5}{x_2} = \frac{x_4}{x_1} + \frac{x_4}{x_2} \qquad (p = 11),$$

(40.74) $$\frac{x_1}{x_4} + \frac{x_1}{x_5} + \frac{x_5}{x_4} = 1 + \frac{x_2}{x_3} + \frac{x_3}{x_2} \qquad (p = 11),$$

(40.75) $$\left(1 + \frac{x_1}{x_2}\right)\left(1 + \frac{x_3}{x_2}\right) = \frac{x_2}{x_4}\left(1 + \frac{x_5}{x_3}\right)\left(1 + \frac{x_5}{x_2}\right) \qquad (p = 11),$$

(40.76) $$\frac{x_4}{x_2}\left(1 + \frac{x_3}{x_2}\right)^2 = \left(1 + \frac{x_2}{x_1}\right)\left(1 + \frac{x_5}{x_2}\right)^2 \qquad (p = 11).$$

These, together with the secondary relations (40.67) can certainly not be independent.

The simplicity of some of these identities leads one to conjecture that the ratios x_M/x_N have coefficients (in their power series expansions) that are of arithmetical significance, and that the identities could then be read off by a comparison of coefficients.

41. Permanent identities. We observe that (40.7) and (40.72) are identical in form, and computation shows that the same equation holds for $p = 13$. Thus we are led to suspect the existence of *permanent identities*, that is, those which are valid for all $p \geq p_0$. We shall now outline and illustrate a method for obtaining such identities.

Fix $p_0 = 2\nu_0 + 1$, and choose only those M, N in (40.3) and (40.31) for which $M + N \leq \nu_0$. Then for all $p \geq p_0$, (M, N) falls under I or IV, according as $N = 2N'$ or $N = p - 2N'$, so $\varepsilon = f_{M,N} = 1$, and (40.3) takes the form

$$(41.1) \qquad x_M + x_{M+N} = 2c \frac{x_M x_{M+N} x_{N'}}{x_s} \cdot \frac{A_M A_{M+N} A_{N'}^2}{A_N A_s^2} \qquad (p \geq p_0).$$

We classify the indices in (41.1) as *permanent* if they do not involve p, *accidental* if they do. Thus, $M, M + N$, and N are always permanent, N' and s are permanent if N is even, accidental if N is odd. In the first case, $N' = N/2$, $s = M + N' = M + N/2$, and in the second, $N' = (p - N)/2$, $s = N' - M = \{p - (2M + N)\}/2$. The vector $\varsigma(M, N)$ may be written in the form

$$(41.11) \qquad \varsigma(M, N) = \xi(M, N) + 2\sigma(M, N),$$

where ξ contains only permanent indices, σ only accidental ones. Thus, if a linear combination $\sum a_r \varsigma(M_r, N_r)$ has the property that $\sum a_r \sigma(M_r, N_r) = 0$, it must have the same value for all $p \geq p_0$, and the corresponding product must have the same form, as a function of the q's and A's, for all $p \geq p_0$. In particular, if $\sum a_r \varsigma(M_r, N_r) = 0$ for $p = p_0$, and if $\sum a_r \sigma(M_r, N_r) = 0$ for $p \geq p_0$, then the product cannot involve the A's, for all $p \geq p_0$, and we have an identity of the type considered in the previous section. Furthermore, since the accidental A's all cancel, so do the accidental q's, which occur with half the frequency, by (41.1). But this means that we have a permanent identity.

To facilitate computation, we make correspond to $\sum a_r \varsigma(M_r, N_r)$ an *indicial vector*, defined as

$$(41.12) \qquad \eta = (d_1, d_3, d_5, \dots),$$

where d_ω is the exponent to which $x_{(p-\omega)/2}$ is raised in the product. Those pairs with N even contribute nothing to d_ω, but for N odd we get $+1$ if $N = \omega$, -1 if $2M + N = \omega$, since $s = (2M + N)'$. Thus

$$(41.13) \qquad d_\omega = \sum_{N_r = \omega} a_r - \sum_{2M_r + N_r = \omega} a_r.$$

The correspondence between η and $\sum a_r \sigma(M_r, N_r)$ is one-to-one; if one is zero, so is the other. Linear combinations of the vector equations correspond to the same linear combinations of their indicial vectors. When $\sum a_r \varsigma(M_r, N_r) = 0$ for $p = p_0$, and $\eta = 0$, the permanent identity may be written down explicitly from (41.1):

$$(41.14) \quad \prod_r (x_{M_r} + x_{M_r+N_r})^{a_r} = \prod_r (x_{M_r} \cdot x_{M_r+N_r})^{a_r} \prod_{N_r=2N'_r} \left(\frac{x_{N'_r}}{x_{M_r+N'_r}}\right)^{a_r}.$$

As an example, take $p_0 = 13$. We get ten vector equations, each accompanied by its indicial vector. For convenience, we omit the ς in $\varsigma(M, N)$.

(41.20) $(3,2) + (2,3) - 2(1,3) = 0$, $\qquad \eta_1 = (1, -1, 2, -1, 0, 0)$,

(41.21) $(1,1) + (1,3) - (2,1) - (2,2) = 0$, $\qquad \eta_2 = 0$,

(41.22) $(1,1) + (1,4) - (2,4) - (1,5) = 0$, $\qquad \eta_3 = (1, -1, -1, 1, 0, 0)$,

(41.23) $(1,1) + (4,1) - (4,2) - (5,1) = 0$, $\qquad \eta_4 = (1, -1, 0, 0, -1, 1)$,

(41.24) $(1,4) + 2(3,3) - (4,1) - 2(1,3) = 0$, $\qquad \eta_5 = (-1, 0, 2, 0, -1, 0)$

(41.25) $(1,1) + (3,3) + (1,4) - (1,5) - (2,1) - (1,3) = 0$,

$$\eta_6 = (0, -1, 1, 1, -1, 0),$$

(41.26) $(1,2) + (2,1) + 2(1,3) + 4(4,1) - 2(1,1) - 6(3,3) = 0$,

$$\eta_7 = (3, -2, -3, 0, 2, 0),$$

(41.27) $(2,3) + (1,1) + 4(3,3) - 3(4,1) - 3(1,3) = 0$,

$$\eta_8 = (-2, 1, 3, -1, -1, 0),$$

(41.28) $(3,1) + 2(3,3) - (1,3) - 2(4,1) = 0$, $\qquad \eta_9 = (-1, 1, 1, -1, 0, 0)$,

(41.29) $(1,1) + 3(3,3) + (5,1) - 3(4,1) - (2,1) - (1,3) = 0$,

$$\eta_{10} = (-2, 1, 2, 0, 0, -1).$$

Upon reducing the set of indicial vectors, we find the following six relations:

(41.31) $\qquad\qquad\qquad \eta_2 = 0,$

(41.32) $\qquad\qquad\qquad \eta_3 + \eta_9 = 0,$

(41.33) $\qquad\qquad\qquad \eta_4 + \eta_{10} = \eta_5,$

(41.34) $\qquad\qquad\qquad \eta_3 + \eta_5 = \eta_6,$

(41.35) $\qquad\qquad\qquad \eta_5 + \eta_9 = \eta_8,$

(41.36) $\qquad\qquad\qquad \eta_6 + \eta_7 + \eta_8 = \eta_1 + \eta_3.$

Using these to direct our combinations of (41.20)–(41.29), we obtain the permanent vector equations:

(41.41) $(1,1) + (1,3) = (2,1) + (2,2)$,

(41.42) $(1,1) + (1,4) + (3,1) + 2(3,3) = (2,4) + (1,5) + (1,3) + 2(4,1)$,

(41.43) $2(1,1) + (3,3) + (1,3) = (4,1) + (4,2) + (2,1) + (1,4)$,

(41.44) $(1,4) + (3,3) + (2,1) = (2,4) + (4,1) + (1,3)$,

(41.45) $(1,1) + (2,3) = (3,1) + (1,4)$,

(41.46) $(1,2) + (4,1) + (2,4) = (1,1) + (3,3) + (3,2)$.

An equivalent but somewhat simpler system is

$$(41.51) \qquad (1,1) + (1,3) = (2,1) + (2,2),$$

$$(41.52) \qquad (1,3) + (1,5) = (2,3) + (4,2),$$

$$(41.53) \qquad (1,1) + (2,3) = (1,4) + (3,1),$$

$$(41.54) \qquad (1,2) + (1,4) = (2,2) + (3,2),$$

$$(41.55) \qquad 2\{(1,2) - (3,2)\} = (4,2) - (2,4),$$

$$(41.56) \qquad (1,4) + (2,1) + (3,3) = (1,3) + (2,4) + (4,1).$$

Using (41.14) and simplifying, we obtain the permanent identities:

$$(41.61) \qquad \frac{x_1}{x_4} + \frac{x_2}{x_1} = \frac{x_3}{x_2} + \frac{x_3}{x_4},$$

$$(41.62) \qquad \frac{x_1}{x_4} + \frac{x_6}{x_1} = \frac{x_5}{x_2}\left(1 + \frac{x_6}{x_4}\right),$$

$$(41.63) \qquad \frac{x_2}{x_1} + \frac{x_5}{x_2} = \frac{x_3}{x_4}\left(1 + \frac{x_5}{x_1}\right),$$

$$(41.64) \qquad \frac{x_1}{x_5} + \frac{x_3}{x_1} = \frac{x_4}{x_2}\left(1 + \frac{x_3}{x_5}\right),$$

$$(41.65) \qquad \frac{x_5}{x_1}\left(1 + \frac{x_1}{x_3}\right)^2\left(1 + \frac{x_2}{x_6}\right) = \frac{x_4}{x_2}\left(1 + \frac{x_5}{x_3}\right)^2\left(1 + \frac{x_4}{x_6}\right),$$

$$(41.66) \qquad \left(1 + \frac{x_5}{x_1}\right)\left(\frac{x_2}{x_3} + \frac{x_3}{x_6}\right) = \left(1 + \frac{x_2}{x_6}\right)\left(\frac{x_5}{x_4} + \frac{x_4}{x_1}\right).$$

These cannot, of course, be algebraically independent.

Every permanent identity implies an identity in two variables. For suppose that

$$(41.7) \qquad\qquad J(x_1, x_2, \dots, x_k) = 0$$

for $p \geq p_0$. Define

$$(41.71) \qquad Y_a(z;q) = \prod_{n \geq 1} \frac{(1 - z^n q^{-2a})(1 - z^{n-1} q^{2a})}{(1 + z^n q^{-2a})(1 + z^{n-1} q^{2a})},$$

where q is (temporarily) fixed, $0 < |q| < 1$, and z is a complex variable, $|z| < 1$. It is clear that Y_a, $a \geq 1$, has no singularities in the circle $C_a : |z| < |q|^{2a}$. Since we may assume that J is a polynomial in the x_a, the function

$$(41.72) \qquad\qquad J^*(z;q) = J(Y_1(z;q), \dots, Y_k(z;q))$$

is regular (in z) in the circle C_k. But J^* has zeros at all the points $z = q^p$, $p \geq p_0$, so it vanishes identically in C_k, and by analytic continuations, in $|z| < 1$. If we now fix z, $|z| < 1$, J^* is meromorphic in q in the region $0 < |q| < \infty$, and vanishes in $0 < |q| < 1$. Hence

$$(41.73) \qquad\qquad J^*(z;q) \equiv 0, \qquad |z| < 1, 0 < |q| < \infty.$$

Now, in the notation of Whittaker and Watson [**47**],

(41.74) $$Y_a(z;q) = i\frac{\vartheta_1(au|\tau)}{\vartheta_2(au|\tau)}, \qquad z = e^{2\pi i\tau}, \quad q = e^{iu}.$$

Thus, if we write $g(u) \equiv g(u|\tau) = Y_1(z;q)$, we have

(41.75) $$J(g(u), g(2u), \ldots, g(ku)) \equiv 0.$$

We have therefore devised a method for generating identities among theta-functions with arguments $u, 2u, \ldots, ku$ for arbitrarily large k.

Incidentally, we have also proved that a permanent identity is valid for all $p \geq 2k + 1$, where k is the highest index appearing. For example, (41.63) and (41.64), proved for $p \geq 13$ originally, hold also for $p = 11$.

Suppose now that we have a permanent identity (41.7). Then, by (41.73), $J^*(q^{2p};q) = 0$. But $Y_a(q^{2p};q)$ is an even function of q, so we may replace q^2 by q throughout. Since

(41.8) $$Y_a(q^p;q^{1/2}) = \prod_{n\geq 1} \frac{(1 - q^{pn-a})(1 - q^{pn-p+a})}{(1 + q^{pn-a})(1 + q^{pn-p+a})} = x_{a'},$$

where, as usual, $a' = a/2$ or $(p - a)/2$, whichever is an integer, it follows that

(41.81) $$J(x_{1'}, x_{2'}, \ldots, x_{k'}) = 0.$$

Thus, every permanent identity is invariant under the substitution $x_a \to x_{a'}$, for each $p \geq 2k + 1$. The resulting identity, however, is not necessarily a permanent one. For example, consider (41.61). For $p = 9$, we have the substitution (142), so we get the new identity

(41.82) $$\frac{x_4}{x_2} + \frac{x_1}{x_4} = \frac{x_3}{x_2} + \frac{x_3}{x_1} \qquad (p = 9).$$

This corresponds to the vector equation

(41.83) $$\varsigma(1, 3) + \varsigma(2, 2) = \varsigma(1, 1) + \varsigma(1, 2),$$

which is verified without too much trouble. The indicial vector for this equation is $\eta = (-1, 2, -1) \neq 0$, so (41.82) is not permanent.

Applying this device with $p = 11$ to (41.63), the substitution being (15342), we get

(41.84) $$x_2(x_1 + x_5)(x_1 + x_3) = x_1(x_2 + x_4)(x_3 + x_5),$$

which corresponds to the permanent vector equation

(41.85) $$\varsigma(1, 4) + \varsigma(1, 2) = \varsigma(2, 2) + \varsigma(3, 2).$$

If we apply the substitution again to (41.84), we find

(41.86) $$x_1(x_3 + x_5)(x_4 + x_5) = x_5(x_1 + x_2)(x_3 + x_4) \qquad (p = 11),$$

which corresponds to

(41.87) $$\varsigma(3, 2) + \varsigma(4, 1) = \varsigma(1, 1) + \varsigma(3, 1).$$

This is not permanent, the indicial vector being $\eta = (-1, 1, 0, 1, -1)$.

In using this transformation, we have had to start with a permanent identity like (41.63), perform the substitution to get (41.84), and identify the latter as being derivable from the vector equation (41.85), *which then had to be verified* by checking the individual vectors $\varsigma(1,4), \varsigma(1,2), \varsigma(2,2), \varsigma(3,2)$, using (40.6) and (40.61). It was necessary to know that (41.85) actually held, in order to use the concept of permanence for (41.84). In all cases so far it has turned out that the vector equation, which was guessed from the transformed identity, is a valid one. We shall now prove that this is always the case.

Let us suppose that we know an identity of the form

$$(41.9) \qquad \prod_r (x_{M_r} + x_{M_r+N_r})^{a_r} = \prod_{i=1}^{\nu} x_i^{b_i},$$

where $0 < M_r, N_r \le \nu$, $M_r + N_r \le \nu$, for a particular $p = 2\nu + 1$. The a_r, b_i are arbitrary integers, except that $\sum a_r = 0$. We suspect, and wish to prove, that the vector identity

$$(41.91) \qquad \varsigma \equiv \sum_r a_r \varsigma(M_r, N_r) = 0$$

holds. For this purpose, we form the product corresponding to ς, using (40.3) and (40.31), and recalling that here $\varepsilon_r = f_r = 1$, $j_r = M_r + N_r$:

$$(41.92) \qquad \prod_r (x_{M_r} + x_{M_r+N_r})^{a_r} = (2c)^{\Sigma a_r} \prod_r \left(\frac{x_M x_j x_{N'}}{x_s} \cdot \frac{A_M A_j A_{N'}^2}{A_N A_s^2} \right)_r^{a_r},$$

the subscript r on the right to be attached to each of the indices within the parentheses. Comparing this with (41.9), and using the fact that $\sum a_r = 0$, we get

$$(41.93) \qquad \prod_{i=1}^{\nu} x_i^{c_i} = \prod_{i=1}^{\nu} A_i^{e_i},$$

say, with $\sum e_i = 0$. Now define

$$U_j \equiv U_j(q) = \prod_{n \ge 1} (1 - q^{pn-j})(1 - q^{p(n-1)+j}),$$

$$V_j \equiv V_j(q) = \prod_{n \ge 1} (1 + q^{pn-j})(1 + q^{p(n-1)+j}).$$

From the definitions of A_i, B_i, and $x_i = B_i/A_i$, we see that (41.93) is equivalent to

$$(41.94) \qquad \prod_{j=1}^{\nu} U_j^{\alpha_j} V_j^{\beta_j} = \left\{ \prod_{n \ge 1} (1 - q^{pn})^2 \right\}^{\Sigma e_i} = 1,$$

with $\{\beta_j\}$ a permutation of $\{-c_i\}$, $\{\alpha_j\}$ a permutation of $\{c_i + e_i\}$. Thus, to show that the components of $\varsigma = (e_1, e_2, \ldots, e_\nu)$ are all zero, it is sufficient to prove that $\alpha_j = \beta_j = 0$, $j = 1, 2, \ldots, \nu$. Now, taking logarithms in (41.94) and expanding each logarithm, we obtain

$$0 = \sum_{j=1}^{\nu} \left[\alpha_j \sum_{n \geq 0} \sum_{k \geq 1} k^{-1} q^{k(pn+j)} + \alpha_j \sum_{n \geq 1} \sum_{k \geq 1} k^{-1} q^{k(pn-j)} \right]$$

$$+ \sum_{j=1}^{\nu} \left[\beta_j \sum_{n \geq 0} \sum_{k \geq 1} (-1)^k k^{-1} q^{k(pn+j)} + \beta_j \sum_{n \geq 1} \sum_{k \geq 1} (-1)^k k^{-1} q^{k(pn-j)} \right].$$

Equating to 0 the coefficient of q^N, $N \geq 1$, yields

$$0 = \sum_{j=1}^{\nu} \alpha_j \sum_{\substack{d \mid N \\ d \equiv \pm j \pmod p}} d + \sum_{j=1}^{\nu} \beta_j \sum_{\substack{d \mid N \\ d \equiv \pm j \pmod p}} (-1)^{N/d} d.$$

For N odd, this may be written

(41.95) $$\sum_{j=1}^{\nu} (\alpha_j - \beta_j) \sum_{\substack{d \mid N \\ d \equiv \pm j \pmod p}} d = 0.$$

Now define the arithmetic function $g(d) = d(\alpha_j - \beta_j)$ if d is odd and $d \equiv \pm j$ $(\bmod\, p)$, $j = 1, 2, \ldots, \nu$, and $g(d) = 0$ otherwise. Then, by (41.95), we may write $\sum_{d \mid N} g(d) = 0$ for all positive integers N. The Möbius inversion formula then shows that $g(d) = 0$ for all d. Now if j is odd, $\alpha_j - \beta_j = j^{-1} g(j) = 0$, and if j is even, $p - j$ is odd, so

$$\alpha_j - \beta_j = (p - j)^{-1} g(p - j) = 0.$$

Thus, for $j = 1, 2, \ldots, \nu$, $\beta_j = \alpha_j$, and the product (41.94) may be written

$$\prod_{j=1}^{\nu} (U_j V_j)^{\alpha_j} = 1.$$

But $U_j(q) V_j(q) = U_j(q^2)$, so

$$\prod_{j=1}^{\nu} U_j^{\alpha_j}(q^2) = 1.$$

Replacing q^2 by q, we get

$$\prod_{j=1}^{\nu} U_j^{\alpha_j} = 1.$$

But this is a product of the same type as (41.94), with $\alpha_j^* = \alpha_j$, $\beta_j^* = 0$. By the preceding argument, $\alpha_j^* = \beta_j^* = 0$, so $\alpha_j = 0$, and finally $\beta_j = \alpha_j = 0$. This completes the proof.

42. Continuation. There are still other means of deriving equations among the x_a for a specific p from a permanent identity, say

(42.1) $$J(x_1, x_2, \ldots, x_k) = 0.$$

As we have seen in §41, this implies

$$(42.11) \qquad J(Y_1(z;q), Y_2(z;q), \ldots, Y_k(z;q)) = 0.$$

Now it is easy to see that for all integers a, r,

$$(42.21) \qquad Y_{p-a}(q^p;q) = -Y_a(q^p;q),$$

$$(42.22) \qquad Y_{p+a}(q^p;q) = Y_a(q^p;q),$$

$$(42.23) \qquad Y_a(q^p;q^r) = Y_{ar}(q^p;q),$$

$$(42.24) \qquad Y_{ap}(q^p;q) = 0.$$

Thus, if we define

$$(42.25) \qquad x_a = Y_a(q^p;q)$$

for all a, we can reduce any identity among the x_a $(-\infty < a < +\infty)$ to the range $1 \le a \le \nu = (p-1)/2$. If we now replace z by q^p and q by q^r in (42.11), we get

$$(42.3) \qquad J(x_r, x_{2r}, x_{3r}, \ldots, x_{kr}) = 0$$

by using (42.23) and (42.25). This is still a permanent identity. We can then use the conversion formulas

$$(42.31) \qquad x_a = \begin{cases} x_b & \text{if } a \equiv b \pmod{p}, \\ -x_b & \text{if } a \equiv -b \pmod{p}, \end{cases}$$

to reduce the arguments in (42.3) to the usual range $1 \le a \le \nu$. The same reduction can also be used for a permanent identity (42.1) with $k > \nu$.

As an example, consider the permanent identity (41.61):

$$(42.41) \qquad \frac{x_1}{x_4} + \frac{x_2}{x_1} = \frac{x_3}{x_2} + \frac{x_3}{x_4}.$$

Under $a \to 2a$, we get

$$\frac{x_2}{x_8} + \frac{x_4}{x_2} = \frac{x_6}{x_4} + \frac{x_6}{x_8},$$

and for $p = 11$, this reduces to

$$(42.42) \qquad \frac{x_4}{x_2} + \frac{x_5}{x_4} = \frac{x_2 + x_5}{x_3}.$$

Under $a \to 3a$, with $p = 11$, we get

$$(42.43) \qquad \frac{x_2}{x_5} + \frac{x_5}{x_3} = \frac{x_2 + x_3}{x_1}.$$

Under $a \to 4a$, with $p = 11$,

$$(42.44) \qquad \frac{x_1}{x_5} + \frac{x_3}{x_4} = \frac{x_1}{x_3} + \frac{x_4}{x_5}.$$

Finally, under $a \to 5a$, with $p = 11$,

$$(42.45) \qquad \frac{x_5}{x_2} + \frac{x_1}{x_5} = \frac{x_4}{x_1} + \frac{x_4}{x_2}.$$

This last can also be obtained by the transformation $a \to a'$, with $1 \to 5 \to 3 \to 4 \to 2 \to 1$.

We can also increase our supply of permanent identities as follows. Start with the permanent identity

(42.5) $x_2(x_3 + x_4)(x_1 + x_5) = x_4(x_1 + x_2)(x_2 + x_5),$

which corresponds to (41.53). For $p \geq 11$, the substitution $a \to a'$ yields

(42.51)
$$x_1(x_2 + x_{(p-3)/2})(x_{(p-5)/2} + x_{(p-1)/2}) = x_2(x_1 + x_{(p-1)/2})(x_1 + x_{(p-5)/2}),$$

corresponding to the vector equation

(42.52) $\varsigma\left(2, \frac{p-7}{2}\right) + \varsigma\left(\frac{p-5}{2}, 2\right) = \varsigma\left(1, \frac{p-3}{2}\right) + \varsigma\left(1, \frac{p-7}{2}\right).$

Now if $p \equiv 3 \pmod 4$, the second indices are all even, so (42.52) is permanent. If $p \equiv 1 \pmod 4$, it turns out that the indicial vector is zero. Thus, if we write $p = 2m + 5$, the vector equation

(42.53) $\varsigma(1, m - 1) + \varsigma(1, m + 1) = \varsigma(2, m - 1) + \varsigma(m, 2)$

is permanent for all $m \geq 3$. For $m = 2$, it reduces to (42.51), which we have already found to be permanent. Hence, for all $m \geq 2$,

(42.54) $x_1(x_2 + x_{m+1})(x_m + x_{m+2}) = x_2(x_1 + x_m)(x_1 + x_{m+2})$

is a permanent identity. Defining

(42.55) $\rho_a = x_a/x_1,$

we have

(42.56) $\dfrac{\rho_{m+1}}{\rho_2} = \dfrac{1 + \rho_m \rho_{m+2}}{\rho_m + \rho_{m+2}}$ $(m \geq 1).$

Every general permanent identity like (42.54), of the form

(42.6) $J(x_1, x_2, \ldots, x_k; x_m, x_{m+1}, \ldots, x_{m+s}) = 0$ $(m \geq m_0),$

implies an identity in three variables, as follows. Since (42.6) is permanent,

(42.61) $J(Y_1(z; q), \ldots, Y_k(z; q); Y_m(z; q), \ldots, Y_{m+s}(z; q)) = 0$ $(m \geq m_0).$

For fixed q and z such that $q^m z^n \neq 1$ for any integers n, m, consider

(42.62) $\Phi(w) = J(Y_1(z; q), \ldots, Y_k(z; q); Y_1(z; w); Y_1(z; qw), \ldots, Y_1(z; q^s w)).$

The only possible singularities of $\Phi(w)$ in $0 < |w| < \infty$ are at the points where

$$w^2 = -q^{-2a} z^n   (-\infty < n < \infty, \ a = 0, 1, \ldots, s).$$

By (42.23) and (42.61), $\Phi(q^m) = 0$ for all $m \geq m_0$. Now, directly from definition $Y_1(z; wz) = Y_1(z; w)$, so $\Phi(wz) = \Phi(w)$. Therefore $\Phi(q^m z^n) = 0$ for $m \geq m_0$, $-\infty < n < \infty$. But the set $\{q^m z^n\}$ is dense in the w-plane, so $\Phi(w)$ vanishes identically. By continuity the same result follows for all q, z. Thus, as in the preceding section, we obtain

(42.63) $J(g(u), \ldots, g(ku); g(w), g(w + u), \ldots, g(w + su)) = 0,$

where $g(u) = i\vartheta_1(u|\tau)/\vartheta_2(u|\tau)$. By the homogeneity of J, we could drop the factor i. Specifically, (42.54) yields

$$(42.64) \qquad g(u)(g(2u) + g(w + u))(g(w) + g(w + 2u))$$
$$= g(2u)(g(u) + g(w))(g(u) + g(w + 2u)).$$

This can undoubtedly be verified by classical means.

We have been able to construct another general permanent vector equation:
(42.65)
$$\varsigma(1, 2) + \varsigma(2, m) + \varsigma(2, m + 2) = \varsigma(m + 2, 2) + \varsigma(3, m) + \varsigma(1, m + 2) \qquad (m \geq 1),$$

which yields, after combination with (43.56),

$$(42.66) \qquad \frac{\rho_m \rho_{m+1}(\rho_2 - \rho_m \rho_{m+1})}{(1 - \rho_m^2)(1 - \rho_{m+1}^2)} = \frac{\rho_2^2 \rho_3}{(1 + \rho_3)(1 - \rho_2^2)}.$$

A curious though trivial consequence of the development in this and the preceding section is that every permanent identity remains valid if we replace x_a by $\tan a\theta$ or $\tanh a\theta$, since $Y_a(0; q) = (1 - q^{2a})/(1 + q^{2a})$. Thus (42.61) becomes

$$(42.67) \qquad \frac{\tan \theta}{\tan 4\theta} + \frac{\tan 2\theta}{\tan \theta} = \frac{\tan 3\theta}{\tan 2\theta} + \frac{\tan 3\theta}{\tan 4\theta}.$$

Similarly, from (42.56) we derive

$$(42.68) \qquad f(q, 2y) = \frac{f(q + y, y) + f(q - y, y)}{1 + f(q + y, y)f(q - y, y)},$$

where $f(q, y) = \tanh y/\tanh q$.

A somewhat different type of modular equation, stemming from equation (10.6), may be deserving of further development. In that equation, put $p = 3$, $r = 1$. Then

$$(42.7) \qquad A(q) \equiv \prod \frac{(1 - q^{3n})^3}{(1 - q^n)} = \sum_{n \geq 0} E_1(6n + 2; 6)q^n$$
$$= \sum_{n \geq 0} E_1(3n + 1; 6)q^n.$$

The odd part of $A(q)$ is

$$\frac{1}{2}(A(q) - A(-q)) = \sum_{n \geq 0} E_1(6n + 4; 6)q^{2n+1}$$
$$= \sum_{n \geq 0} E_1(3n + 2; 6)q^{2n+1}$$
$$= \sum_{n \geq 0} E_1(6n + 2; 6)q^{4n+1} + \sum_{n \geq 0} E_1(6n + 5; 6)q^{4n+3}$$
$$= \sum_{n \geq 0} E_1(3n + 1; 6)q^{4n+1},$$

since $E_1(6n + 5; 6) = 0$. Hence

$$(42.71) \qquad 2qA(q^4) = A(q) - A(-q).$$

This leads easily to the modular equation of degree 3 in the form

(42.72) $$1 + (l^3/k)^{1/4} = l'^3/k.$$

Similarly, if we put $p = 5$, $r = 1$ and 3 in (10.6), we get

(42.8) $$B(q) \equiv \prod_{n \geq 1} \frac{(1 - q^{5n})^2}{(1 - q^{5n-1})(1 - q^{5n-4})} = \sum_{n \geq 0} E_1(5n + 2; 10)q^n,$$

(42.81) $$C(q) \equiv \prod_{n \geq 1} \frac{(1 - q^{5n})^2}{(1 - q^{5n-2})(1 - q^{5n-3})} = \sum_{n \geq 0} E_3(5n + 3; 10)q^n.$$

Let us define

$$C_1(q) = \sum_{n \geq 0} E_3(5n + 4; 10)q^n,$$

$$C_2(q) = \sum_{n \geq 0} E_3(5n + 2; 10)q^n,$$

$$C_3(q) = \sum_{n \geq 0} E_3(5n + 1; 10)q^n.$$

It follows from the arithmetic properties of $E_3(n; 10)$ that

$$C(q) - C(-q) = 2qC_1(q^2),$$
$$C_1(q) + C_1(-q) = 2C_2(q^2),$$
$$C_2(q) + C_2(-q) = 2C_3(q^2),$$
$$C_3(q) = qC(q^2).$$

From these we obtain, with $\beta = \exp(\pi i/4)$,

(42.82) $$8q^9 C(q^{16}) = \sum_{k=0}^{7} \beta^{-k} C(\beta^k q).$$

Similarly,

(42.83) $$8q^6 B(q^{16}) = \sum_{k=0}^{7} \beta^{2k} B(\beta^k q).$$

It seems clear that equations of the type (42.82) and (42.83) must exist for every odd p.

Notes

Modular equations have been much studied recently. Of course they appeared early in the study of the theory of elliptic functions. Russell [15] provides an extensive account of early work, and G. H. Hardy gives a readable introduction in Chapter 12 of Ramanujan [10]. Certain types of modular equations played a central role in Watson's proof of Ramanujan's conjectures for powers of 5 and 7 [16, 3, 12, 13].

Recently the power of modern computer algebra packages like MACSYMA has provided access to the actual expressions for some of the classical equations that were not amenable to hand calculation [11]. Consequently the constructive and elementary methods of Fine may be of interest in subsequent constructive developments.

Many of the results given by Fine (e.g. (42.82) and (42.83)) are reminiscent of the types of results found by Ramanujan and elucidated by Birch [7], Bressoud [9], and Rangachari [14]. Surprisingly such seemingly esoteric modular equations have also become important in statistical mechanics (see Baxter [4], and [2]).

Finally it is especially important to call attention to two major recent projects. First is the outstanding computational advances made by J. and P. Borwein that are most vividly epitomized by their calculation of tens of millions of the digits in the decimal expansion of π. Their work is beautifully described in [8]. Indeed the first five chapters of [8] are in much the same spirit as this book. Second there is the extensive account of Ramanujan's work on modular equations done by B. Berndt [1, 5, 6]. Both [5] and [6] will appear in Berndt's third volume of Ramanujan's Notebooks published by Springer.

REFERENCES

1. G. Almkvist and B. Berndt, *Gauss, Landen, Ramanujan, the arithmetic-geometric mean, ellipses, π, and the Ladies Diary*, MAA Monthly (to appear).

2. G. E. Andrews, R. J. Baxter, and P. J. Forrester, *Eight-vertex SOS model and generalized Rogers-Ramanujan-type identities*, J. Statist. Physics **35** (1984), 193–266.

3. A. O. L. Atkin, *Proof of a conjecture of Ramanujan*, Glasgow Math. J. **8** (1967), 14–32.

4. R. J. Baxter, *Exactly solved models in statistical mechanics*, Academic Press, London, 1982.

5. B. C. Berndt, *Chapter 19 of Ramanujan's second notebook, modular equations of degrees 3, 5, and 7 and associated theta function identities*, Ramanujan's Notebooks, Part III, Springer-Verlag (to appear)

6. _____, *Chapter 20 of Ramanujan's second notebook, modular equations of higher and composite degrees*, Ramanujan's Notebooks Part III, Springer-Verlag (to appear)

7. B. J. Birch, *A look back at Ramanujan's notebooks*, Proc. Cambridge Philos. Soc. **78** (1975), 73–79.

8. J. M. Borwein and P. B. Borwein, *Pi and the AGM*, John Wiley, New York, 1987.

9. D. M. Bressoud, *Some identities involving Rogers-Ramanujan-type functions*, J. London Math. Soc. **16** (1977), 9–18.

10. G. H. Hardy, *Ramanujan*, Cambridge Univ. Press, 1940; reprinted by Chelsea, New York.

11. E. Kaltofen and N. Yui, *On the modular equation of order 11*, Proc. 1984 MACSYMA Users Conf., General Electric, Schenectady, 1984.

12. M. Knopp, *Modular functions in analytic number theory*, Markham, Chicago, 1970.

13. J. Lehner, *Lectures on Modular forms*, Nat. Bur. Standards Appl. Math. Series, no. 61, Washington, D.C., 1969.

14. S. Rangachari, *Ramanujan identities and icosahedral groups* (to appear).

15. R. Russell, *On modular equations*, Proc. London Math. Soc. **21** (1890), 351–393.

16. G. N. Watson, *Ramanujans Vermütung über Zerfällungsanzahlen*, J. Reine und Angew. Math. **179** (1938), 97–128.

Bibliography

1. George E. Andrews, *The theory of partitions*, Encyclopedia of Mathematics and its Applications, Vol. 2, Addison-Wesley, Reading, Mass., 1976.

2. ___, *q-Series: Their development and application in analysis, number theory, combinatorics, physics and computer algebra*, CBMS Regional Conf. Ser. Math., No. 66, Amer. Math. Soc., Providence, R.I., 1986.

3. A. O. L. Atkin and P. Swinnerton-Dyer, *Some properties of partitions*, Proc. London Math. Soc. (3) **4** (1954), 84–106.

4. F. C. Auluck, *On some new types of partitions associated with generalized Ferrers graphs*, Proc. Cambridge Philos. Soc. **47** (1951), 679–686.

5. W. N. Bailey, *Generalized hypergeometric series*, Cambridge Tracts in Math., Cambridge Univ. Press, Cambridge-New York, 1935, pp. 65-72.

6. ___, *A further note on two of Ramanujan's formulae*, Quart. J. Math. Oxford Ser. (2) **3** (1952), 158–160.

7. E. T. Bell, *Algebraic arithmetic*, Amer. Math. Soc. Colloq. Publ., Vol. 7, New York, 1927.

8. L. Carlitz, *Note on some partition formulae*, Quart. J. Math. Oxford Ser. (2) **4** (1953), 168–172.

9. A. Cauchy, *Oeuvres* (1) VIII, 42–50.

10. Th. Clausen, J. Reine Angew. Math. **3** (1828), 95.

11. L. E. Dickson, *History of the theory of numbers*, G. E. Stechert and Co., New York, 1934.

12. L. A. Dragonette, *Some asymptotic formulae for the mock theta series of Ramanujan*, Trans. Amer. Math. Soc. **72** (3) (1952), 474–500.

13. F. J. Dyson, *Some guesses in the theory of partitions*, Eureka, Feb. 1944, 10–15.

14. L. Euler, *Introductio in analysin infinitorum*, Lausanne **1** (1748).

15. ___, *Observ. anal. de combinationibus*, Comm. Acad. Petrop. **13** (1741–3, 1951), 64–93.

16. N. J. Fine, *Some new results on partitions*, Proc. Nat. Acad. Sci. U.S.A. **34**(12) (1948), 616–618.

17. ___, *On a system of modular functions connected with the Ramanujan identities*, Tôhoku Math. J. **8**(2) (1956), 149–164.

18. C. F. Gauss, *Werke*, Göttingen, 1876.

19. E. Grosswald, *Representations of integers as sums of squares*, Springer-Verlag, 1985.

20. W. Hahn, *Über Orthogonalpolynome, die q-differenzengleichungen genügen*, Math. Nachr. **2** (1949), 4–34.

21. ___, *Über Polynome, die gleichzeitig zwei verschiedenen Orthogonalsystemen angehören*, Math. Nachr. **2** (1949), 263–278.

22. ___, *Beiträge zur Theorie der Heineschen Reihen...*, Math. Nachr. **2** (1949), 340–379.

23. G. H. Hardy, P. V. Seshu Aiyar, and B. M. Wilson, *Collected papers of Srinivasa Ramanujan*, Cambridge Univ. Press, 1927.

24. G. H. Hardy, *Ramanujan*, Cambridge Univ. Press, 1940.

25. G. H. Hardy and E. M. Wright, *Introduction to the theory of numbers*, Oxford Univ. Press, 1945.

26. E. Hecke, *Lectures on Dirichlet series*, Princeton Univ. Press, 1936.

27. E. Heine, *Theorie der Kügelfunktionen*, Berlin, 1878, pp. 97–125.

28. C. G. J. Jacobi, *Fundamenta nova theoriae functionum ellipticarum*, Regiomonti, fratrum Bornträger, 1829.

29. J. Liouville, *Sur quelques formules générales qui peuvent être utiles dans la théorie des nombres*, Jour. de Math. (2) (1858–1865).

30. P. A. MacMahon, *Combinatory analysis*, Cambridge Univ. Press, 1915.

31. L. J. Mordell, *On Mr. Ramanujan's empirical expansions of modular functions*, Proc. Cambridge Philos. Soc. **19** (1919), 117–124.

32. T. Nagell, *Introduction to number theory*, John Wiley and Sons, Inc., New York, 1951.

33. P. S. Nazimoff, *Applications of the theory of elliptic functions to the theory of numbers*, Moscow, 1884; transl. by A. E. Ross, Chicago, 1928.

34. E. Netto, *Lehrbuch der Combinatorik*, B. C. Teubner, Leipzig and Berlin, 1927.

35. L. J. Rogers, *On a three-fold symmetry in the elements of Heine's series*, Proc. London Math. Soc. **24** (1893), 171–179.

36. ____, *On the expansion of some infinite products*, Proc. London Math. Soc. **24** (1893), 337–352.

37. ____, *Second memoir on the expansion of certain infinite products*, Proc. London Math. Soc. **25** (1894), 318–343.

38. ____, *Third memoir on the expansion of certain infinite products*, Proc. London Math. Soc. **26** (1895), 15–32.

39. D. B. Sears, *Transformations of basic hypergeometric functions of special type*, Proc. London Math. Soc. (2) **52** (1951), 467–483.

40. ____, *On the transformation theory of basic hypergeometric functions*, Proc. London Math. Soc. (2) **53** (1951), 158–180.

41. ____, *Transformations of basic hypergeometric functions of any order*, Proc. London Math. Soc. (2) **53** (1951), 181–191.

42. G. W. Starcher, *On identities arising from solutions of q-difference equations and some interpretations in number theory*, Amer. J. Math. **53** (4) (1930), 801–816.

43. J. J. Sylvester, *A constructive theory of partitions*, Coll. Math. Papers, Vol. IV, Cambridge Univ. Press, 1912, pp. 1–83; (reprinted by Chelsea, N.Y., 1974).

44. G. Szegö, *Ein beitrag zur Theorie der Thetafunktionen*, Sitz. der Berlin Akad., 1926, pp. 242–252.

45. G. N. Watson, *The final problem: An account of the mock-theta functions*, J. London Math. Soc. **11** (1936), 55–80.

46. H. Weber, *Lehrbuch der Algebra*, Vol. 3, Chelsea, New York, 1908.

47. E. T. Whittaker and G. N. Watson, *Modern analysis*, Cambridge Univ. Press, 1943.